Structural Failure Analysis and Prediction Methods for Aerospace Vehicles and Structures

Edited by

Sook-Ying Ho

Defence Science and Technology Organisation
Edinburgh, SA, Australia

CONTENTS

FOREWORD

An important engineering problem in structural design is the evaluation of structural integrity and reliability. It is well known that structural strength may be degraded during its design life due to mechanical and/or chemical aging. Depending on the structural design, material type, service loading, and environmental conditions, the cause and degree of strength degradation due to the different aging mechanisms will vary. One of the common causes of strength degradation is crack development in the structure. When cracks occur, the effects of crack size and rate of growth on the fracture resistance of the material and the remaining strength and life of the structures need to be determined.

Reliable performance of a structure in critical applications depends on ensuring that the structure in service satisfies the conditions assumed in design and life prediction analyses. Reliability assurance requires evaluation of the stress state in the structure and the material's strength allowable corresponding to a given failure criterion for a given loading condition and the availability of nondestructive testing and evaluation techniques to characterize discrete cracks according to their location, size, and orientation. This leads to an improved assessment of the potential criticality of individual flaws.

This book addresses the important topics mentioned in the above paragraphs. It covers the areas of mechanical fatigue analysis of aircraft structures under constant amplitude and spectrum loading conditions, acoustic fatigue of secondary aircraft structures, aerodynamic heating, structural integrity and service life prediction methodology of solid rocket motors, and structural health monitoring of solid rocket motors. Based on a systematic and a blended experimental and analytical approach together with extensive practical experience, methodologies were developed to (1) predict fatigue crack growth behavior and service life of aircraft structures and (2) assess the structural integrity and service life of solid rocket motors. In addition, the applicability as well as the limitations of the various methods are pointed out and examples of their practical use are given. Though the subject matter that appears in this book has been previously published by the authors elsewhere, this is the first time that the information has been presented in a consolidated manner. The implementation of these advanced technologies can increase the reliability of aircraft and rocket structures as well as significantly reduce structural repair and replacement costs. This book will prove to be of considerable value to the field of aerospace engineering both in academia and in industry. Furthermore, I believe its value will extend beyond the immediate intended reader to engineers in the shipping, automobile, tire, and construction industries who are interested in the evaluation of integrity and service life of structures.

Dr. Chi Tsieh (Jimmy) Liu
Principal Research Scientist (retired)
Propulsion Directorate
Air Force Research Laboratory
USA

PREFACE

This book deals with structural failure - induced by mechanical, aerodynamic, acoustic and aero-thermal loads - of modern aerospace vehicles, in particular high-speed aircraft, solid propellant rocket systems and hypersonic flight vehicles. Structural integrity and failure prediction (essential for the design and service life assessment) of modern aerospace structures and flight vehicles have significant challenges due to the increasingly more demanding mission requirements and the use of non-traditional materials, such as non-metallic composites, in their construction. The growing pressure for high reliability and longer operational service life adds to the problem.

Prediction of the complex loading environment (such as aerodynamic heating seen in high-speed flights, thermal and acoustic operational and/or storage loads), and constitutive / fracture models can adequately describe the non-linear behaviour exhibited by advanced alloys and composite materials. These are some critical areas in predicting the non-linear structural response and failure analysis of modern aerospace flight vehicles. It is also important to use appropriate prediction methodologies for a given type of vehicle. For example, the methods for predicting the structural integrity and failure of solid propellant rocket systems are different to the methods used for aircraft, largely due to the non-linear viscoelastic behaviour of solid propellants and the different operational / environmental loading conditions. Another example is acoustic fatigue, a major damage phenomenon, that has to be considered in modern fighter aircraft but not for expendable solid rocket boosters.

This book has seven chapters, describing different key issues /challenges, emerging technologies, experimental methods and analytical techniques for predicting / assessing structural integrity and failure for the three selected types of modern aerospace vehicles: high-speed aircraft, solid propellant rocket systems and hypersonic flight vehicles.

This book represents what is believed to be a current need of the aerospace vehicle design engineer / scientist and structural analyst, arising from the necessity for more accurate and appropriate methods of (1) structural and failure analysis and (2) predicting the complex operational / environmental loading conditions for different types of modern aerospace vehicles and structures. The chapters are written by experts from aerospace / defence research organisations, industry and academia in the fields of structural mechanics and dynamics of aircraft, rocket and hypersonic systems. It will serve as a useful reference document containing specialist knowledge on appropriate prediction methodologies for a given circumstance and experimental data acquired from multi-national collaborative programs. It is aimed at providing (1) a practical professional tool for all engineers, scientists and program managers working in the various disciplines of aerospace structures and vehicles, and (2) an effective reference text for graduate students and academia in the aeronautical and astronautical sciences.

In the first chapter, *Barter, Molent and Wanhill* present an overview of current and proposed methods for aircraft fatigue life testing and an innovative lead crack fatigue lifing framework developed at the Air Vehicles Division of DSTO for fatigue life assessment of high performance metallic airframe structures. Fatigue crack growth (FCG) and quantitative fractography (QF) data based on many years of detailed inspection and analysis of fatigue cracks in airframe materials and structures, ranging from coupon to full-scale fatigue tests (including actual components from in-service F/A-18A/B and F-111 aircraft), are presented. A step-by-step procedure for determining the lead crack at any given life and location and examples of lead crack FCG estimates (from multiple cracks or if there is no information on initial discontinuities) are presented.

Although good designs to reduce a structure's susceptibility to sonic fatigue are now available, structural failures resulting from sonic fatigue remain a significant issue for some fighter aircraft where relatively overall high sound pressure levels were found to exist on the external surface of the nacelle skin. In Chapter 2, *Callinan, Wang, Galea and Sanderson* describe acoustic fatigue analysis and repair of sonic fatigue damaged panels. Details on a state-of-the-art generic design procedure (using closed form solutions based on crack bridging theory) and a real-life case study of a design of adhesively bonded patch repair to prevent acoustic fatigue cracking on the aft fuselage of the F/A-18 aircraft are presented. Data of sound pressure levels measured on the external surface of the aircraft are included.

Aero-thermal-structural analysis is a fundamental area for the design and failure analysis of hypersonic vehicles and structures. Unlike low-speed flight vehicles where pressure and skin friction have a significant effect on the aerodynamic forces and moments and hence, the flight stability of the vehicle, aerodynamic heating is the predominant structural load in the high-speed flight regime. Some consequences of aerodynamic heating include melting and /or deformation of structural components and aerothermoelasticity effects. Chapter 3 (*Ho*) is devoted to describing the multidisciplinary aspects of aero-thermal and structural dynamic analysis of hypersonic vehicles. Example applications of the numerical techniques, including aero-thermal-structural analysis of a real-life hypersonic flight vehicle, are also presented.

New approaches for predicting fatigue crack growth in aircraft structures and components under flight spectrum loading are described by *Zhuang and Molent* in Chapter 4. Predictions of fatigue crack growth (FCG) are made directly based on the flight spectrum fatigue crack growth data measured by quantitative fractography (QF) for several common aircraft structural materials under various fighter aircraft flight spectra. The new approach provides much more reliable assessment of the crack growth rates than the predictions deduced from traditional constant amplitude crack growth testing and cycle-by-cycle algorithmns. The QF process identifies the morphological features on a fatigue crack surface attributed to the applied specific spectrum loading sequences. Predictions of fatigue crack growth for different stress concentration factors at different load levels are described, with the aim to develop better prediction of FCG life for airframes containing different geometric discontinuities.

The design and failure criteria for modern aerospace vehicles and advanced structures have increased in complexity over the past years, due to the more complex loading environment, geometries and new materials used in the construction. It is, therefore, desirable to have generic fracture models that are suitable for all material types, stress-strain states and loading conditions, and hence appropriate to a wide range of applications. New three-dimensional predictive models of various non-linear fatigue crack growth phenomena that are suitable for advanced structures under complex loading conditions (such as, constant amplitude, retardation following an overload cycle) are presented in the chapter by *Kotousov and Codrington*. These models, based on refined plate theory, can take into account thickness effect, material non-linearity and variable loading cycle effects. Possible applications of these models include damage tolerance design and analysis of fracture-critical aircraft engine components, as well as service life analysis of solid propellant rocket systems.

In the past two decades, there have been considerable advances in non-destructive evaluation (NDE) methods for detecting fracture, etc. and structural health monitoring of aircraft and rocket propulsion systems. Smart sensor technology can offer enormous benefits in terms of cost savings and increased operational safety and reliability of aerospace structures and vehicles. Chapter 6 (*Ho*) describes state-of-the-art miniature stress sensor technology for monitoring the thermal stresses and ignition pressurization loads in a generic research end-burning solid rocket motor. This study was part of a larger international collaborative effort carried out from 1988 to 2002 to validate the instrumentation and analytical stress analysis and service life prediction methodologies for solid composite rocket motors. Stress sensor selection and installation, data reduction and the use of the instrumented rocket motor data for (1) structural health monitoring, (2) validating theoretical viscoelastic models and stress / strain predictions from finite element analysis and (3) input into probabilistic service life prediction models are discussed in this chapter. Other NDE methods for detecting structural failure in solid composite rocket motors are also discussed.

The application and role of structural integrity assessment (computational and experimental methods) to service life prediction and vulnerability / safety assessment of solid rockets are presented in chapter 7 (*Ho*). Accurate methods for analyzing the stress-strain response, failure criteria and environmental / operational loads are essential in order to deal with the increasingly more stringent requirements in performance, safety, reliability and cost of these systems. The chapter illustrates the use of computational techniques, such as the finite element method, and appropriate failure criteria for structural integrity assessment, and an example solution procedure for a real-life rocket. A method for determining high-strain-rate mechanical properties using the modified Hopkinson Bar technique and the development of appropriate constitutive models are also described.

Sook-Ying Ho
Defence Science and Technology Organisation
Edinburgh, SA, Australia

LIST OF CONTRIBUTORS

Simon A. Barter
Air Vehicles Division
Defence Science and Technology Organisation
Melbourne
Australia
E-mail: simon.barter@dsto.defence.gov.au

Richard J. Callinan
Air Vehicles Division
Defence Science and Technology Organisation
Melbourne
Australia
E-mail: Richard.Callinan@dsto.defence.gov.au

John Codrington
School of Mechanical Engineering
The University of Adelaide
Australia.
John.codrington@adelaide.edu.au

Steve Galea
Air Vehicles Division
Defence Science and Technology Organisation
Melbourne
Australia
E-mail: Steve.Galea@dsto.defence.gov.au

Sook-Ying Ho
Weapons Systems Division
Defence Science and Technology Organisation
Edinburgh
Australia
E-mails: sookying.ho@dsto.defence.gov.au
        sookying.ho@adelaide.edu.au

Andrei Kotousov
School of Mechanical Engineering
The University of Adelaide
Australia
E-mail: andrei.kotousov@adelaide.edu.au

Lorrie Molent
Air Vehicles Division
Defence Science and Technology Organisation
Melbourne
Australia
E-mail: Lorrie.Molent@dsto.defence.gov.au

Stephen Sanderson
Air Vehicles Division
Defence Science and Technology Organisation
Melbourne

Australia
E-mail: Stephen.Sanderson@dsto.defence.gov.au

Chun Wang
Sir Lawrence Wackett Aerospace Centre
School of Aerospace
Mechanical and Manufacturing Engineering
RMIT, Victoria
Australia
E-mail: chun.wang@rmit.edu.au

Russell Wanhill
Aerospace Vehicles Division
National Aerospace Laboratory
NLR, Amsterdam
The Netherlands
E-mail: Wanhill@nlr.nl

Wyman Zhuang
Air Vehicles Division
Defence Science and Technology Organisation
Melbourne
Australia
E-mail: wyman.zhuang@dsto.defence.gov.au

NOMENCLATURE

The following is a list of those symbols used in the text to define the main parameters. Due to the many disciplines represented in this text and the preference to adhere to common usage wherever possible, some symbols are used to represent more than one parameter. The correct parameter should be determined from the context in which the symbol is used. Some symbols used for special purposes are not listed but will be defined in the text.

A	Crack area, plate area
C	Cure fluid weight correction;
	Damping matrix;
	Viscous damping matrix
C_f	Coefficient for crack growth law;
	Skin friction coefficient
C_D	Drag coefficient
C_L	Lift coefficient
C_M	Pitching-moment coefficient
D	Bending stiffness of plate;
	Cumulative Damage fraction
D_f	Transducer stress disturbance factor
E	Young's modulus;
	Flux vector in x direction
$\{E\}, \{F\}$	Stress vectors components
E'	Real part of complex Young's modulus
E''	Imaginary part of complex Young's modulus
$E(\xi)$	Relaxation modulus
Ex_t	Transducer excitation power
F	Geometry factor;
	Empirical heating factor;
	Flux vector in y direction
$\{F\}$	Applied load vector
G	Shear modulus
G'	Real part of complex shear modulus
H	Enthalpy
H^*	Reference enthalpy
H_2	Enthalpy behind a normal shock wave
H_{st}	Stagnation enthalpy
H_R	Boundary layer recovery enthalpy
H_w	Wall enthalpy
H_r	Hysteresis ratio
I	Second moment of area
I_f	Transducer-propellant interaction factor
I_t	Total bending stiffness
I_{tr}	Real part of total bending stiffness
J	J integral or crack extension force;
	Mechanical equivalent of heat (778 ft lb/Btu)
J_2	Deviatoric stress invariant

L_x	Length of plate in x direction
L_y	Length of plate in y direction
K	Stress intensity factor;
	Conductivity matrix;
	Stiffness matrix
K_p	Peak stress intensity factor
K_{min}	Minimum stress intensity factor
K_{mean}	Mean stress intensity factor
K_{max}	Maximum stress intensity factor
K_{rms}	Root mean square stress intensity factor
K_m	Membrane stress intensity factor
K_b	Bending stress intensity factor
K_m^{rms}	Root mean square membrane stress intensity factor
K_b^{rms}	Root mean square bending stress intensity factor
$\overline{K}$	Stress intensity factor due to application of unit pressure on structure
K_{rms-u}	Root mean square stress intensity factor un-repaired
K_{rms-r}	Root mean square stress intensity factor repaired
K_{th}	Threshold stress intensity factor for crack growth
K_{tn}	Net-section stress concentration factor
K_{eff}	Effective stress intensity factor
$(K_{max})_{rms}$	Maximum rms stress intensity factor
K_T	Stress intensity factor due to thermal stresses
K_m^T	Stress intensity factor due to thermal membrane stresses
K_b^T	Stress intensity factor due to thermal bending stresses
M	Mass (inertia) matrix;
	Mach number
M_∞	Freestream Mach number
P_1	Static pressure in front of a shock wave
P_2	Static pressure behind a shock wave
Pr_w	Prandtl number
R	Nose or leading edge radius;
	Ratio between minimum and maximum stress (constant amplitude tests)
Re	Reynolds number
S	Reynolds Analogy factor;
	Solar radiation input
S_e	Transducer sensitivity at temperature T
S_o	Power spectral density of the response
S_I	Power spectral density of the excitation
$S_I(f_0)$	Power spectral density of the excitation at frequency f_0
S_{ij}	Deviatoric stress component
St	Stanton number (ratio of actual heating rate to the total potential heating rate)
T	Temperature
$\{T\}$	Vector of nodal temperatures
T_{aw}	Adiabatic wall temperature
T_e	Temperature at edge of boundary layer
T_{st}	Stagnation (or total) temperature, also T_t or T_o

T_w	Wall temperature
ΔT	Change from cure to operating temperature
T_c	Cure temperature
T_{oper}	Operating temperature
U	Solution (conservation variables) vector
$\{U\}$	Displacement vector
U_e	Velocity of flow at edge of boundary layer
W	Geometry factor for frequency
W_i	Input strain energy density
W_{rc}	Recoverable strain energy density
V_2	Velocity behind a shock wave
Y	Geometric factor
$ZO(T)$	Transducer zero offset at temperature T
a	Half crack length;
	Speed of sound
a_T	Time-temperature shift factor
c	Damping constant
d_1	Distance between centroid of section and centre of plate
d_3	Distance between centroid of section and centre of constraining layer
e	Internal energy
e_t	Total energy
f	Vibration frequency of plate
f_m	Membrane component of the
	Normalized stress intensity factor
f_b	Bending component of the normalized stress intensity factor
f_o	First resonant frequency of the un-repaired plate
f_c	First resonant frequency of the repaired plate
$f(t)$	Externally imposed forcing function (load vector)
Δf	Frequency range
g	Softening function;
	Gravitational conversion factor (32.17 lbm ft/lb s^2)
g_{opt}	Optimum value of shear parameter
g_m	Geometric shear parameter
g'	Real part complex geometric shear parameter
$g*$	Complex shear parameter
h	Thickness of plate;
	Heat transfer coefficient
k_f	The wave number
k_x	Modal wave number in x direction
k_y	Modal wave number in y direction
m_t	Total mass per unit length
m	Fundamental mode in x direction
n	Fundamental mode in y direction
p_{rms}	rms fluctuating pressure
p_∞	Freestream pressure
q	Dynamic pressure

$\dot{q}$	Heat flux
r	Recovery factor
s	Stiffness ratio
$s(t)$	Time-dependent randomly varying stress or strain
t	Thickness;
	Time
w	Transverse displacement of plate
α	Thermal coefficient of expansion;
	Weibull modulus value;
	Angle of attack
β	Fraction of peak load that crack opens;
	Weibull modulus value;
κ	Thermal conductivity
ε	Strain
ε_b	Bending strain
ε_m	Membrane strain
$\dot{\varepsilon}$	Strain-rate
$\bar{\xi}$	Reduced time
η	Loss factor;
	Viscosity, also μ
η_s	Loss factor for damped plate
η_c	Loss factor for viscoelastic damping layer
$\eta_{s,max}$	Optimum loss factor for 3 layer plate
η_t	Non-core loss factor
$\eta_{DC,m}$	Modal loss factor
ν	Poissons ratio for plate
σ_b	Far field bending stress
σ_m	Far field membrane stress
σ_{xx}^T	Thermal stress in x direction
σ_{yy}^T	Thermal stress in y direction
σ_{xx}, σ_{yy}	Normal stress components
σ_p	Measured stress from stress sensor
τ	Wall or skin thickness
τ_{xy}	Shear stress
τ_p	Plastic shear stress for adhesive
τ_{peak}	Peak shear stress for adhesive
ρ	Density
ρ_e	Density of flow at edge of boundary layer
ρ_w	Density of wall material
θ	Angle from stagnation point
ω	Natural frequency
ω_M	m[th] resonant frequency
ϕ	Mode shape
w	Transverse displacement
w_o	Transverse displacement at centre of panel

$x(t)$	Nodal displacement vector
$\dot{x}(t)$	Nodal velocity vector
$\ddot{x}(t)$	Nodal acceleration vector
γ	Specific heat ratio
Ω	Non-dimensional frequency ratio
CLD	Constrained layer damping
OASPL	Overall Sound Pressure Level
Pdf	Probability density function
PSD	Power Spectral Density
RT	Room temperature
SPL	Sound pressure level
FCG	Fatigue crack growth

Fatigue Life Assessment for High Performance Metallic Airframe Structures – An Innovative Practical Approach

S.A. Barter[1], L. Molent[1] and R.J.H. Wanhill[2]

[1]*Air Vehicles Division, Defence Science and Technology Organisation DSTO, Melbourne, Australia and* [2]*Aerospace Vehicles Division, National Aerospace Laboratory NLR, Amsterdam, the Netherlands*

Abstract: A new fatigue lifing approach has been developed by the DSTO for high performance metallic airframe structures. This approach is based on years of inspection and analysis of fatigue cracks in many airframe components and specimens, and is now an important additional method of determining fatigue lives for aircraft in the Royal Australian Air Force fleet. Like the earlier Damage Tolerance approach developed by the United States Air Force, the DSTO approach assumes that fatigue cracking begins as soon as an aircraft enters service. However, there are major and fundamental differences. The Damage Tolerance approach specifies a "standard" set of initial crack/flaw sizes based on Non-Destructive Inspection capabilities, and the important period of early (short) crack growth is estimated from back-extrapolation of long crack growth data. On the other hand, the DSTO approach uses initial crack/flaw sizes representative of small, fatigue-initiating discontinuities in the materials and structural components, and the crack growth lives are estimated from actual data for short-to-long cracks growing from these discontinuities. Furthermore, these data, particularly for *lead cracks*, are characterized by exponential crack growth behaviour. Owing to this general characteristic, the DSTO approach can use growth data for the lead crack to provide reasonable lower-bound estimates of component crack growth lives. Scatter factors based on engineering judgement are then applied to these estimates to determine the maximum allowable service life.

1. INTRODUCTION

1.1. Fatigue Life Testing for Metallic Airframes

Accurate estimation of the fatigue lives of metallic airframes still presents challenges, particularly for high performance aircraft. There is always a demand for lighter structures with reduced manufacturing and operating costs. This leads to relatively highly stressed and efficient designs where fatigue cracking can arise at features such as shallow radii at the junction of flanges, webs and stiffeners, as well as at holes and tight radii. As a consequence, there are usually many areas that need to be assessed for their fatigue lives, and many potential locations at which cracking may occur in service.

It is well-known that metal fatigue is a complex phenomenon depending on many parameters. These include the material characteristics (mechanical properties, microstructure, inherent discontinuities, e.g. constituent particles), surface treatments and finishes, the component and structural geometries, dynamic load histories, and the environment.

Nevertheless, engineering fatigue design relies in the first instance on baseline coupon tests to assess the many locations identified as susceptible to cracking. The coupons may be loaded by constant amplitude (CA) or representative variable amplitude (VA) load histories, and they may try to represent some feature of a built-up structure. The results of these coupon tests are averaged to give an indication of the structural life for a production aircraft. However, there are significant limitations to this approach:

1) Experience has shown that in high performance aircraft, the components have many features with the potential to crack, and that each of these features is typical of a single type of "representative" coupon. Hence a component's average indicated life is equivalent to only the shortest average life from tests on several types of coupons.

2) Even when the most critical feature of a component has been identified and assessed by coupon testing, the coupons are rarely fully representative, notably with respect to the surface treatments and finishes required for production aircraft. This is important because the commencement of fatigue cracking is primarily surface-influenced and therefore greatly dependent on small surface

discontinuities inherent to component production, as well as any surface-connected discontinuities inherent to the material.

These limitations are addressed by other means. One way, which is mandatory for all modern aircraft, is to test actual components, part of the structure or even the full airframe, thereby including the effects of component geometry and production. Another way is to improve coupon testing by making the coupons optimally representative of the most fatigue-critical details, e.g. by applying surface treatments and finishes used in component production. This may seem obvious, but it is sometimes neglected or overlooked.

1.2. Fatigue Lifing Methods

There are several possible methods for assessing aircraft fatigue lives. Table **1** gives a survey of actual and proposed methods, including the new approach developed by the DSTO and now being used for the Royal Australian Air Force (RAAF) fleet. Some comments on these methods are made here.

Table 1: Survey of current and proposed methods for aircraft fatigue life assessment

• **Stress – life (S – N)**
   ◦ fatigue limits, S_e; unnotched and notched (K_t); constant amplitude (CA) data
   ◦ modifications to S_e
   ◦ mean stress effects (R)
   ◦ linear damage rule, also for variable R
   ◦ scatter factors

• **Strain – life (ε – N)**
   ◦ strain − life equation, unnotched data, R = -1
   ◦ cyclic stress – strain curve analysis
   ◦ rainflow cycle counting (closed hysteresis loops)
   ◦ stress – strain at critical location (notch analysis)
   ◦ mean stress effects (R) via equivalent strain equations, leading to equivalent strain amplitudes
   ◦ damage accumulation rule

• **Damage Tolerance (DT)**
   ◦ specified equivalent initial flaw sizes (EIFS) based on NDI capabilities
   ◦ back-extrapolation of long crack growth data to derive short crack growth
   ◦ LEFM long crack growth models (non-interaction, yield zone, crack opening, strip yield) to derive variable amplitude (VA) crack growth from constant amplitude (CA) data
   ◦ possible use of crack opening model for short cracks (FASTRAN); differences in long and short crack thresholds need to be included
   ◦ mainly deterministic: stochastic approach becoming accepted

• **Holistic approach: proposed**
   ◦ fatigue initiation mechanisms (also as functions of notch stress concentrations, K_t)
   ◦ fatigue initiation lives (S – N and/or ε – N assessments)
   ◦ evaluation and selection of marker load strategies for Quantitative Fractography (QF) of short-to-long crack growth
   ◦ actual short-to-long long crack growth using marker loads and QF
   ◦ establishment, validation and choice of appropriate crack growth models and "laws"
   ◦ deterministic ("upper bound") and stochastic approaches
   ◦ environmental effects, notably corrosion

• **DSTO approach: proposed and implemented for the RAAF**
   ◦ actual initial discontinuity/flaw sizes and their equivalent pre-crack sizes (EPS)
   ◦ actual short-to-long crack growth data using Quantitative Fractography (QF)
   ◦ data compilations to establish empirical relationships describing crack growth behaviour
   ◦ deterministic ("upper bound") estimates of *lead crack* growth
   ◦ scatter factors

Stress – life (S – N)

This method has been used extensively in the past. It relies on simple cumulative damage rules to obtain "safe" fatigue lives under VA load histories. These safe lives are then factored down using scatter factors based on engineering judgement.

Strain – life (ε – N)

This is a common method used to track fatigue damage in aircraft fleets, and there are several Original Equipment Manufacturer (OEM) and publicly available tools for doing this. (There is always a need to know the effectiveness of these tools and what their weaknesses are).

Damage Tolerance (DT)

This method was originally developed by the United States Air Force (USAF) [1] and is currently used for most military aircraft. However, there are major problems. The equivalent initial flaw size (EIFS) concept is dubious, especially with respect to the minimum crack dimensions that have to be assumed. There is now a consensus that the EIFS requirements can lead to predicted fatigue crack growth (FCG) lives that are (much) too conservative. This has led to setting up the Research and Technology Organisation (RTO) working group AVT-125 "Future Airframe Lifing Methodologies" within the NATO community.

Another serious problem is that the predicted early crack growth behaviour is highly questionable, since it is derived from back-extrapolation of (a) VA long crack growth data or (b) VA growth curves derived from long crack CA data, with both methods using analytical models "tuned" to long crack growth behaviour.

Holistic Approach

This is an ambitious programme, see Fig. **1**, and is intended for all types of aircraft. The inclusion of corrosion effects is especially noteworthy. The project sponsor is the Canadian National Research council (NRC).

DSTO approach

This method is much less ambitious than the HOLSIP programme, but is nevertheless directly applicable to high performance aircraft. The DSTO approach is the main subject of this chapter and will be discussed in detail in the remaining sections.

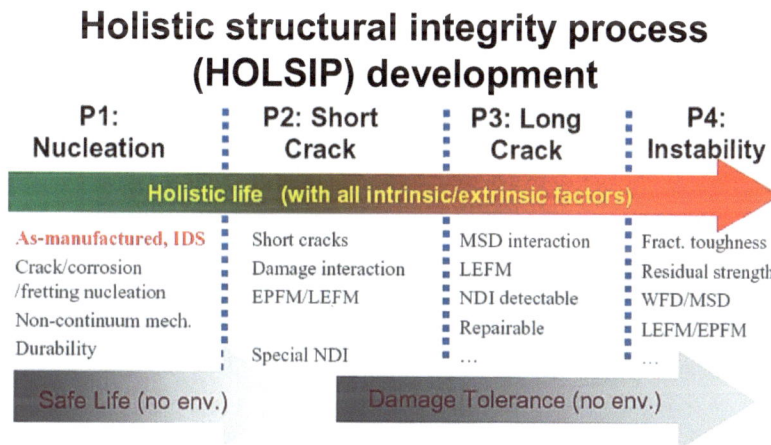

Figure 1: Overview of the HOListic Structural Integrity Process (HOLSIP) programme [2]. IDS = Initial Defect State, EPFM = Elastic Plastic Fracture Mechanics, LEFM = Linear Elastic Fracture Mechanics, MSD = Multi Sight Damage, NDI = Non-Destructive Inspection and WFD = Widespread Fatigue Damage.

2. DSTO APPROACH: BACKGROUND

The DSTO has decades of experience in fatigue testing of metallic airframe materials as coupons, components and full-scale structures; and also much experience in examining fatigue cracks in service aircraft. Detailed inspections and analysis of the fatigue cracks, notably with the aid of quantitative fractography (QF), have shown that most of the largest and naturally-initiating cracks grew in an approximately exponential manner [3-6]. This behaviour has also been observed by others, e.g. [7-9].

The QF observations covered crack sizes from a few micrometres up to many millimetres, and showed that the cracks originated from small discontinuities inherent to the material and from component production. For high performance aircraft these observations were - and are - essential, since most of the fatigue life is spent in growing relatively small cracks.

The QF data have made it possible to (a) characterize the crack-generating discontinuities and their populations, (b) account for variability in small crack FCG behaviour, a notorious problem that is difficult or impossible to tackle in any other way, and (c) predict total fatigue lives from larger or smaller discontinuities.

Furthermore, the largest cracks were found to start growing shortly after the beginning of testing, or shortly after the aircraft entered service.

Based on all these observations, the DSTO has developed a service component lifing approach called the *lead crack fatigue lifing framework* [10]. This framework is now an important additional method for the determination of component fatigue lives for several types of aircraft in the RAAF fleet.

3. LEAD CRACKS

As indicated in section **2**, there have been many observations that approximately exponential FCG is a common behaviour for the largest and naturally-initiating cracks in coupons, components, full-scale structures and service aircraft, e.g. [3-6].

These are the *lead cracks*, i.e. those leading to failure. At this point it is important to note that the lead cracks come from a typical population of cracks. They are not cracks growing from the rare and exceptional discontinuities classed as "rogue flaws" [11].

3.1. Lead Crack Characteristics

The lead cracks have the following characteristics:

1)   They start to grow shortly after testing begins or an aircraft enters service.

2)   Subject to several conditions, they grow approximately exponentially with time, i.e. FCG may be represented by an equation of the form:

$$a = a_0 e^{\lambda N}$$

where a is the crack size at time N, a_0 is the initial crack size, and λ is a constant that includes the geometrical factor β, see point (3) below.

The conditions for approximately exponential FCG are:

a)   The crack does not grow into an area with a significant thickness change, particularly if the crack length/depth is small compared to the coupon or component thicknesses or widths.

b)   The crack is not unloaded, either (i) by the cracked area losing stiffness and shedding load; or because it grows (ii) towards a neutral axis due to bending loads, or (iii) away from an externally-induced stress concentration.

c)   The crack does not encounter a significantly changing stress field by growing into or from an area containing residual stresses.

d)   FCG is not retarded by infrequent very high loads.

e)   The small fraction of FCG life influenced by quasi-static fracture close to final failure is ignored.

These conditions might seem very restrictive, but the DSTO's FCG analyses have shown that approximately exponential and conservative fits can be made for most lead cracks [3-6, 10-12]. Within the bounds of these conditions, the DSTO's analyses have led to the following deductions:

3)   The usually important geometrical factor β (which depends on the ratios of the crack length and shape to the width and thickness of the coupon or component) does not influence FCG as much as might be expected. For low K_t features most of the life is spent when the crack is physically small, so β does not

change much. However, even when a crack starts at an open hole and the calculated β changes rapidly, the lead cracks still appear to grow approximately exponentially. The reason or reasons for the small influence of β require further research.

4) Typical initial discontinuity sizes are about equivalent to a 0.01mm deep crack [13, 14]. In other words, an equivalent pre-crack sizes (EPS) of 0.01mm is a good starting point for FCG assessment. N.B: this EPS value is well below the smallest EIFS values to be assumed in the USAF DT approach [1].

5) Cracks may also grow exponentially within residual stress fields, although the exponents will be influenced by these stress fields [10, 14, 15].

6) If (very) high loads occur fairly frequently in the load history, then the average FCG may still have an exponential trend [3].

7) The metallic materials used in high performance aircraft have typical critical crack depths of about 10mm [13, 4]. Although final failure may be at larger crack sizes, it is usually found that significant quasi-static fracture preceded failure, meaning that the components or structures would not have survived their safety-margin residual strength requirements [10].

3.2. Examples of Exponential FCG and Lead Cracks

Fig. **2** gives examples of approximately exponential FCG curves extending over 2-3 orders of magnitude in crack size. The data were obtained from QF measurements of crack growth in the lower skin of an F-111 wing removed from service and tested under flight-by-flight block loading [16]. The FCG rates (gradients of the FCG curves) were mostly similar, even though the cracks occurred at numerous span-wise and chord-wise locations and covered two-thirds of the wingspan. This means that despite variations in geometrical details and locations, similar FCG rates pertained under similar loading conditions. This trend has also been observed for other aircraft structure [4].

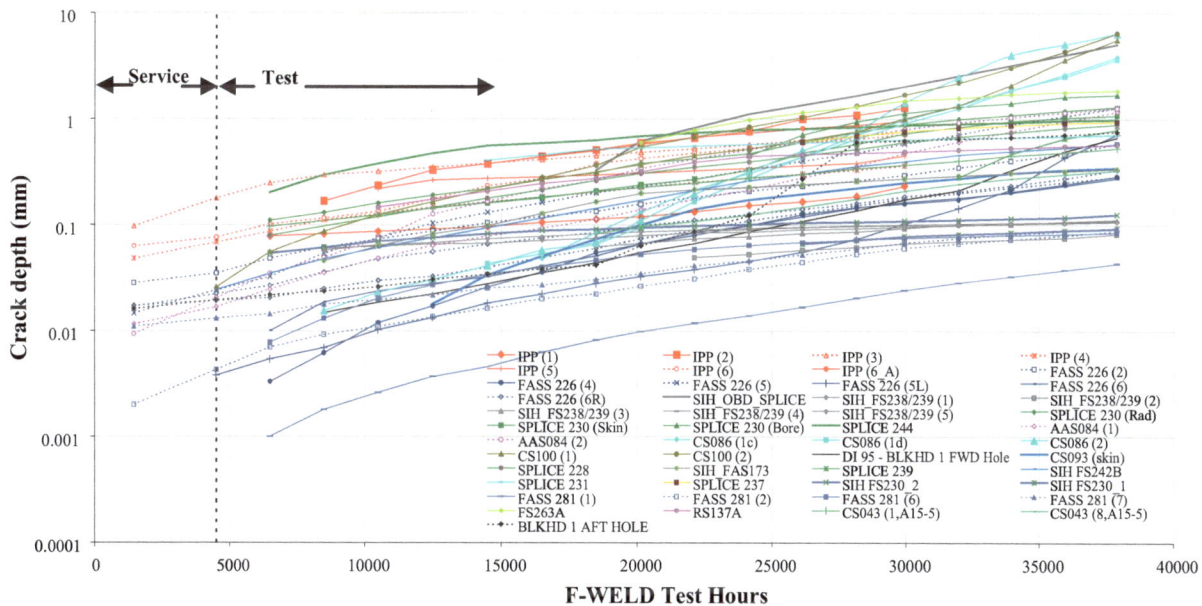

Figure 2: FCG curves from different locations in the AA2024-T851 aluminium alloy wing lower skin of an F-111 test article removed from service [16]. CS = central spar, BLKHD = bulkhead, FASS = Forward Auxiliary Spar Station inches, IPP = Inner Pivoting Pylon, RS = Rear Spar.

Besides demonstrating the exponential trends and similar FCG rates, the log crack size *versus* linear life plot in Fig. **2** has a couple of additional advantages:

1) The FCG behaviour of small cracks is more clearly seen than on a double-linear plot.

2) Given mostly similar FCG rates, it is seen that the major source of scatter is the initial discontinuity/crack size. This demonstrates the importance of obtaining good estimates of EPS values.

From the example FCG curves in Fig. **2** and the above discussion it may be seen that simplified but reasonably accurate FCG life estimates can be derived as follows:

- Assume immediate in-service exponential FCG from initial discontinuities.

- Select several locations and areas.

- Choose a number of initial discontinuity/crack sizes (EPS) and final crack sizes characteristic of these locations and areas.

- Choose characteristic exponent values (or possibly one overall value if the FCG rates are similar) in combination with initial discontinuity/crack sizes (EPS) and final crack sizes to estimate FCG lives for these locations and areas.

Comparison of these estimates will enable determining which of the cracks is the lead crack at any given life, and also its location. For example, Fig. **2** shows that at 20,000 test hours the lead crack was at SPLICE 244, but at 35,000 test hours the lead crack was at CS086. The reasons for this change are (a) the larger initial crack depth for the SPLICE 244 location and (b) the higher FCG rates for the crack at the CS086 location, whereby this crack eventually overtook the SPLICE 244 crack at approximately 28,000 test hours.

4. FATIGUE CRACK INITIATION

4.1. Ideal *Versus* Realistic Conditions

Under more or less ideal conditions, typically for carefully prepared specimens tested in laboratory air, there can be a significant period of fatigue-induced microstructural damage before fatigue crack initiation (or "nucleation"). The eminent fatigue researcher D.W. Hoeppner prefers nucleation, since the term "initiation" has often been - and still is - used imprecisely, without regard to the physical process or processes that preceded crack formation.

Be that as it may, the commencement of fatigue cracking is primarily surface-influenced and therefore greatly dependent on the surface quality. Production aircraft components and structures have many sources of surface or near-surface discontinuities capable of causing fatigue cracking. These include various forms of machining damage (scratches, grooves, burrs, small tears and nicks); etch pits from surface treatments (pickling, anodising); porosity, especially in thick aluminium alloy plate and castings; and in the case of aluminium alloys and steels, constituent particles that may themselves be cracked. Titanium alloys are a special case, but they too can have material discontinuities, though very rarely [17].

Fig. **3** shows some examples of discontinuities, illustrating their variety. More are illustrated in an extensive report [10]. This report also includes a review of fatigue crack initiation in aluminium alloy components and structures, since these materials are the most widely used in metallic airframes.

Although the discontinuities are mostly very small - of the order of 0.01mm in depth [4, 13] - they can cause fatigue cracking very quickly in highly stressed coupons, components and structures. This has been shown by DSTO studies [3, 4, 6, 10, 18] and others [19, 20] and an example is given in subsection **4.2**. Furthermore, structures with many critical features, e.g. fastener and bolt holes, or large areas under high stresses, are susceptible to multiple crack initiation. One or more of these cracks will lead the others to become the critical cracks that determine the FCG life.

4.2. Example of Fatigue Cracking from Discontinuities

As mentioned above, small discontinuities can initiate fatigue cracking very quickly at high stress levels. It was also mentioned earlier that the commencement of fatigue cracking primarily occurs at the surface. To illustrate these points, and their interrelation, Fig. **4** shows QF data for flight simulation FCG from surface-connected and (slightly) subsurface discontinuities in highly stressed aluminium alloy coupons. Most of the

FCG-initiating discontinuities were smaller than 0.05mm and crack growth began effectively immediately when they initiated from the surface. However, for coupons KS1G3 and KS1G66 the FCG-initiating discontinuities were subsurface and there were apparent delays of 10,000 and 26,000 simulated flight hours (SFH) before crack growth began. The reason for these apparent delays is that subsurface crack growth would have occurred *in vacuo* at much slower rates than in air [21]. Once the cracks became surface-connected their FCG rates became similar to those for other coupons tested at the same stress levels.

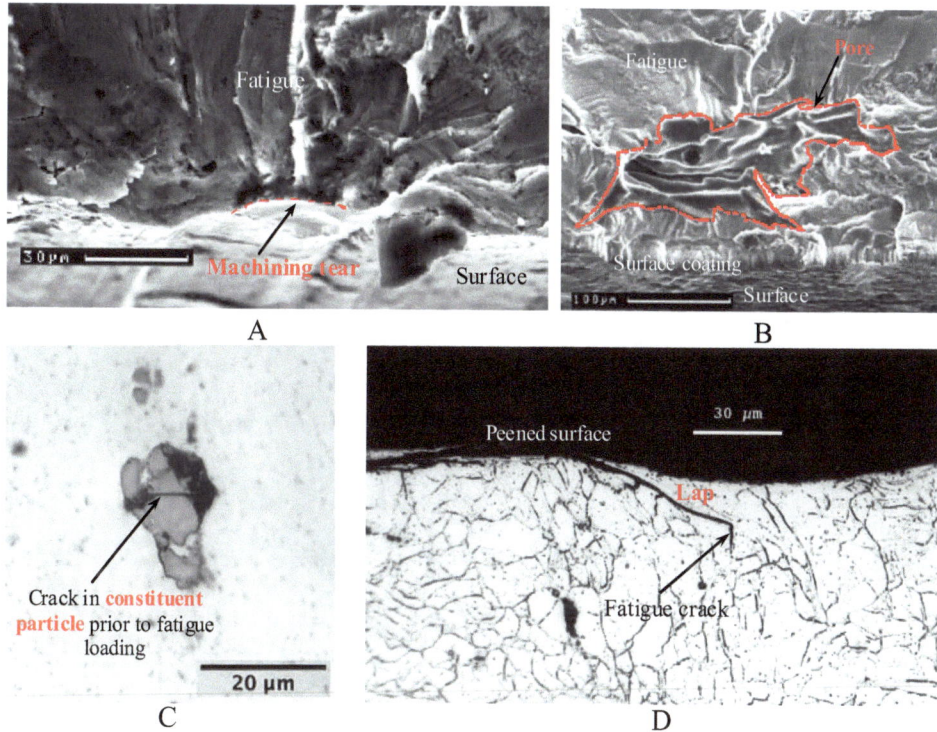

Figure 3: Some examples of discontinuities that can exist in a metallic airframe at the time it enters service. Each discontinuity will act, or has acted, as an effective crack starter, reducing fatigue crack formation to a few effective load cycles.

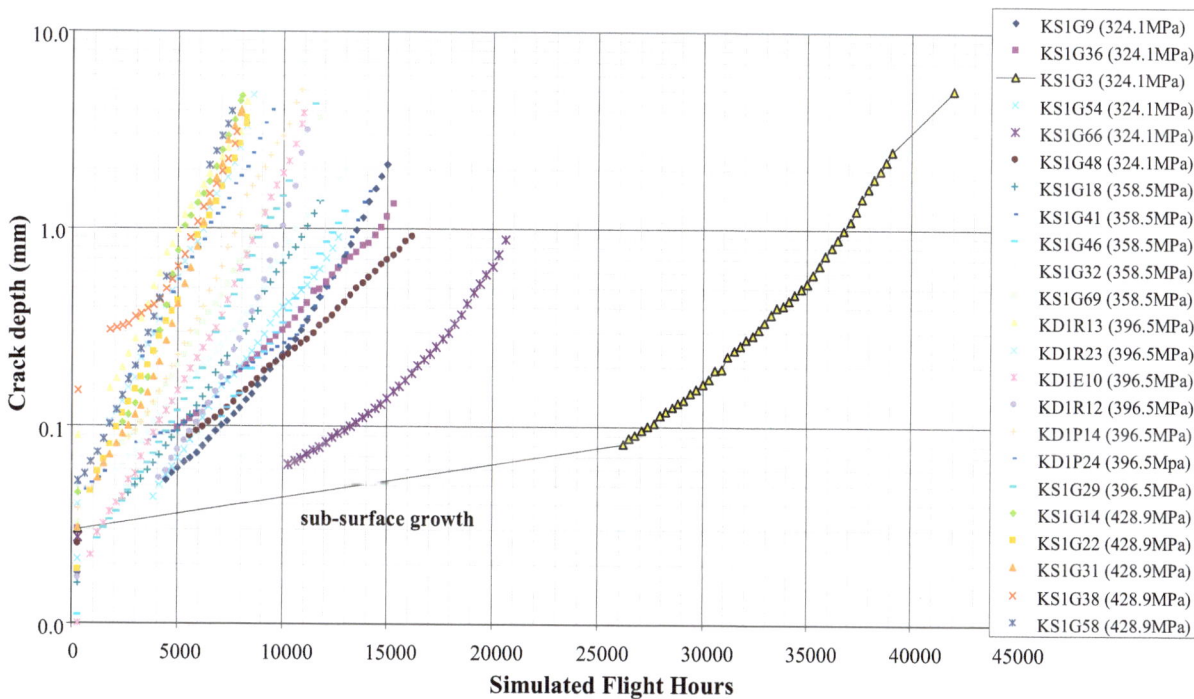

Figure 4: FCG data for highly stressed AA7050-T7451 aluminium alloy coupons tested under combat aircraft flight-by-flight block loading at four stress levels. One block represents about 300 airframe hours and each data point represents the crack growth increment per simulated flight hour.

It must be stated that subsurface fatigue crack initiation is highly exceptional in actual components and structures unless they have undergone surface treatments, e.g. shot peening, to enhance the fatigue resistance. Although such treatments are not relied upon during fatigue design, shot peening is a favoured remedial action for areas found to have insufficient life.

5. FATIGUE CRACK GROWTH

5.1. Approximately Exponential FCG

The observation of exponential or approximately exponential FCG has a long history, going back to the 1950s e.g. [22]. This behaviour is described by the following simple relationships:

$$a = a_0 e^{\lambda N} \tag{1}$$

$$\ell n(a) = \lambda N + \ell n(a_0) \tag{2}$$

These relationships mean that the FCG data appear to be well-represented by straight lines on plots of ℓn (log) crack size *versus* life and ℓn (log) FCG rate *versus* ℓn (log) crack size. Figures 2 and 4 are examples of the first type of plot. Fig. **5** is an example of the second type.

Several points are to be noted about the example in Fig. **5**:

1) The stress level dependence in Fig. **4** is reflected in Fig. **5**.

2) Trend lines for each set of stress level data have slopes reasonably close to 1 *over three orders of magnitude* for both the FCG rates and crack depths. This means that there is a wide range in which the FCG rates are approximately proportional to the crack depths and hence indicate approximately exponential FCG. The wide range of this approximation is a robust affirmation of its applicability and usefulness.

3) There is data scatter with occasional outliers. Some of the scatter may come from QF measurement difficulties, but local material differences can also play a role, especially when the cracks are small and extend through only a few grains [23]. This point is illustrated by Fig. **6**, which is a fracture surface detail from a small fatigue crack in a large-grained aluminium alloy. This detail shows significant local variability in the FCG rate for a block of CA R = 0.7 cycles applied between blocks of 5000 CA R = 0.1 cycles, all with the same maximum loads.

Note that although differentiation to obtain Fig. **5** reveals data scatter, comparison with Fig. **4** shows that the log crack size *versus* linear life plots for each crack are well-behaved. This is a common observation, and other clear examples are given in Ref. [24]. The differentiation to obtain FCG rates accentuates relatively small differences in the progression of crack growth. This can be useful for detailed analyses of crack growth but is a hindrance for predicting FCG lives from crack growth rate data. However, the DSTO approach avoids this problem by using the log crack size *versus* linear life plots to predict the FCG lives. This is shown in section **6**.

5.2. A Note About FCG "Laws"

The approximately exponential FCG behaviour of lead cracks is not a "law" but an empirical relationship, albeit one that is now well-established for VA load histories representative of those experienced by high performance aircraft [10].

There are *no* FCG laws, although many empirical relationships are referred to as such. This was pointed out more than 30 years ago by D.W. Hoeppner and W.E. Krupp [25], who listed some 33 "laws". However, the proliferation of FCG laws continues. There are now at least 186 (D.W. Hoeppner, Dec. 20, 2009).

Most of the laws refer only to CA loading, often using the well-known Paris relationship [26] as a starting point. Also, none do more than implicitly include environmental effects, which have significant influences on FCG, e.g. Refs. [27, 29], especially under VA loading [30, 33].

In short, given the appropriate circumstances one crack growth "law" or relationship is as good as, or better than another. This view guides our assertion of the practical usefulness of the approximately exponential FCG behaviour of lead cracks in aircraft components and structures. This usefulness will be demonstrated next.

6. LEAD CRACK FATIGUE LIFING FRAMEWORK

6.1 Framework

The lead crack fatigue lifing concept has been developed on a framework of observations to calculate virtual test points from full-scale and large component tests, unanticipated service cracking, and sometimes from cracking in representative coupons. The key elements of the framework are (see Section 3):

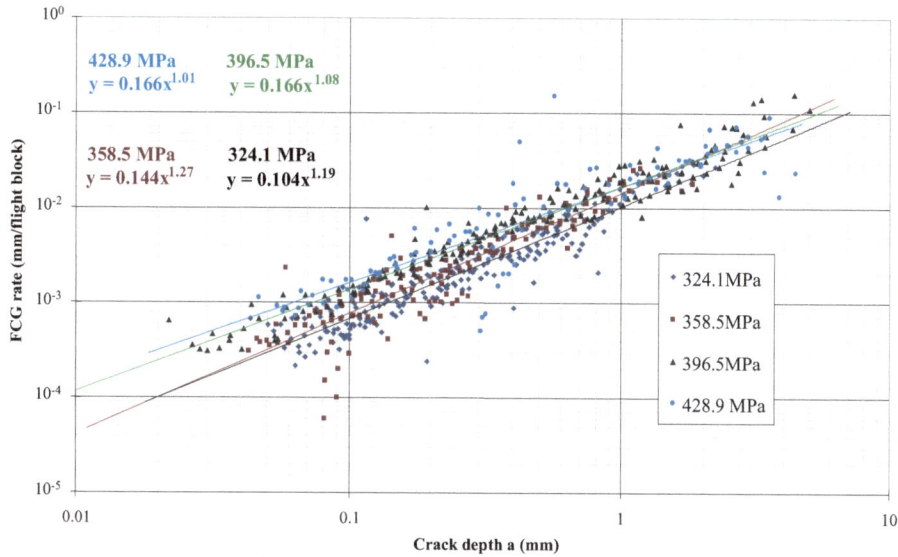

Figure 5: FCG rate data from Fig. 4.

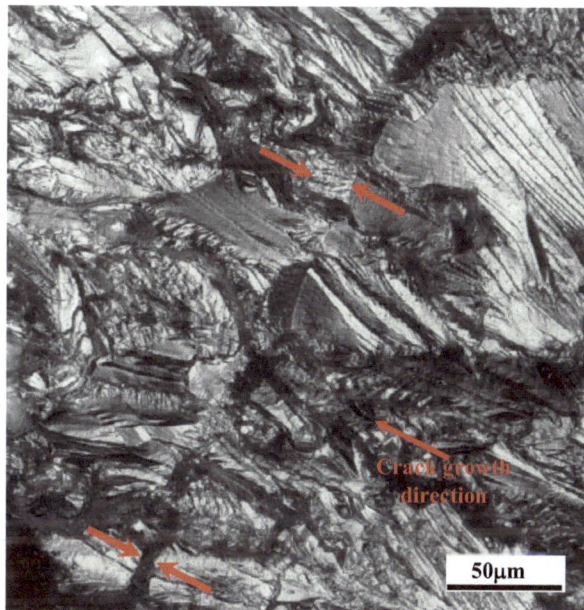

Figure 6: An example of material-related FCG rate variability from the fracture surface of an 7050-T7451 aluminium alloy specimen at about 1mm in crack depth. The band from lower left to top right was formed by a block of CA R = 0.7 cycles applied between blocks of 5000 CA R = 0.1 cycles, all with the same maximum load. Note the width variation of the band at the two arrowed positions.

1)   Lead cracks start to grow shortly after testing begins or the aircraft is introduced into service and subjected to flight loads;

2)   Lead cracks grow *approximately* exponentially with load history;

3)   Lead cracks start growing from material production discontinuities; and

4)   The small fraction of FCG life influenced by quasi-static fracture close to final failure is insignificant.

6.2. Methodology

Referring to Fig. **2**, a step-by-step procedure was presented for determining the lead crack at any given life and location and subsequently calculating its safe life. The procedure, based on the framework described above, contains the following elements:

a)   The assumption of immediate in-service exponential FCG from initial material discontinuities;

b)   An accurate measure of the initial discontinuities (in the form of EPSs) for the locations and areas to be assessed;

c)   An accurate measure of the final crack size characteristics (at end of cycling or failure) for the locations and areas to be assessed;

d)   A measure of the characteristic exponents for FCG from the EPSs to the lead crack of the locations of interest;

e)   An estimate of the FCG lives of the lead cracks which then leads to an estimate of the virtual test points for each location and area;

f)   The pooling of the estimated lives for each location within each area to calculate the average life of the location; and

g)   The application of an appropriate SF[1] to the average life to determine the safe life of the location.

This procedure is a general one. It is feasible and straightforward if there are numerous FCG data from a fully representative full-scale fatigue test. This may not be the case, for several reasons:

1)   An obvious reason is that QF FCG data may be unavailable or unobtainable. These data would have to be acquired from separate programmes of coupon and component testing.

2)   Structural modifications to ensure the survival of the test article result in uncertainties as to the service lives in the modified locations and areas.

3)   The test load history may be unrepresentative in some locations and areas owing to test rig/structural assembly limitations and differing load histories in service.

4)   Unanticipated service cracking problems, i.e. premature cracking and cracks in locations and areas deemed to have been non-critical during prior analysis and testing.

5)   Reasons (2) – (4) require re-assessment of the FCG lives and establishment of new and additional virtual test points for fleet action. These reasons and their accompanying scenarios need a framework of rules for obtaining reliably conservative predictions of FCG lives. This framework is set out in subsection 6.2.

The details of using the methodology are often complex, since it may be necessary to assess and provide robust and comparable lead crack FCG data from different sources [12].

6.3. The Framework Application

The application of the framework may be set out in several steps:

1)   Choice of initial discontinuity and crack sizes characteristic of the locations and areas to be assessed. Characteristic discontinuities can be converted to equivalent pre-crack sizes (EPS) using a well-established empirical expression due to Murakami and Endo [34].

If there is no information on initial discontinuities, e.g. when a crack has been blended out during a full-scale fatigue test, then EPS estimates may be possible by back-extrapolating FCG data for cracks in the same area or location. An example is given in subsection 7.2.

---

[1] Engineering experience, as far as the appropriate selection of a SF is concerned is codified in various design standards.

2)   Choice of final crack sizes characteristic of the locations and areas to be assessed. These crack sizes are preferably the a_{RS} values demonstrated from the RS requirement of 1.2DLL. Or else the critical crack sizes $a_{cr} = a_{RS}$, when further crack growth would have been rapid and structural repairs or modifications were made. Alternatively, the a_{RS} values may be obtained analytically.

3)   Choice of characteristic exponents for fatigue crack growth from the EPS. This step should provide reliably conservative estimates of FCG lives. Hence, much care is needed in deriving and choosing the exponents. The optimum situation is when QF data includes early crack growth from the initial discontinuities. This is not always possible, obviously in the case where a crack has been blended out, or if QF cannot analyse early FCG for a particular load history.

However, the crack sizes at fatigue test end-points and in components retired from service can be used with the EPS to make initial estimates of the exponents. These estimates can be adjusted using FCG data from QF-amenable test load histories applied to (a) similar cracks in test coupons or components and (b) *the same* cracks in retired components.

4)   Combine the previous steps, together with the assumption of immediate in-service fatigue crack growth, to estimate the FCG lives of the lead cracks and their virtual test points for each location and area.

5)   Pool the estimates for each location within each area. This step requires that FCG data and crack growth plots obtained under different circumstances, notably under different load histories, be converted to a common fatigue life timeframe. For the F/A-18 in RAAF service, from which the example data in Fig. **7** has been drawn, this timeframe is the fatigue life expended index (FLEI), which is the estimated *equivalent* service hours at which an aircraft has accumulated the same damage as the F/A-18 fatigue test article known as FT55 (the results of an earlier test, ST16, are also included in this example), divided by the SF, where SF is a scatter factor based on engineering experience.

Further explanation of the relation between the FLEI and design target life of the airframe (AFHRS) is necessary here. For the example given in Fig. **7**, the design target life AFHRS is 6000 hours and the SF is 2.8, so that the required *equivalent* number of full-scale test hours was 16,800. The safe life limit of any aircraft is then reached when the FLEI = 1 = 6000 *equivalent* hours[2].

This rather involved procedure is necessary because individual aircraft will experience differing service load histories, such that a common fatigue life timeframe - the FLEI - must be used to assess the fatigue life expenditure of the fleet. As mentioned above, this common fatigue life timeframe is also required for pooling the lead crack FCG estimates. It should be noted that the in-service calculated FLEI of an individual aircraft is designed to be suitably conservative (see Fig. **7**).

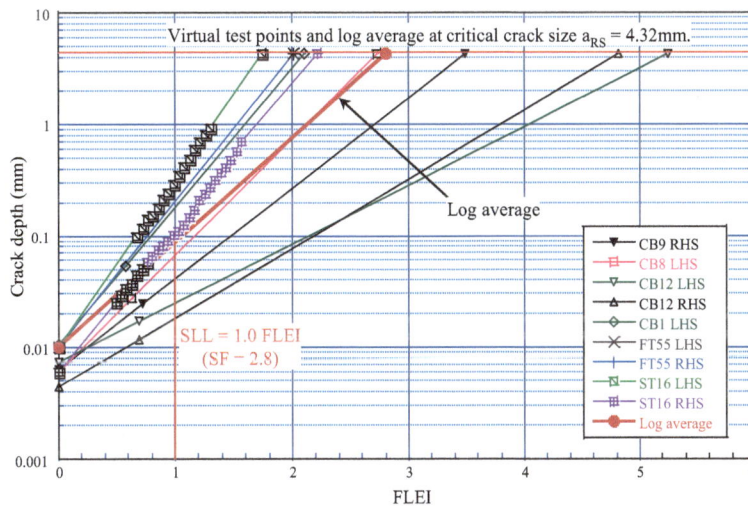

Figure 7: Pooled estimates of exponential FCG for several lead cracks, their virtual test points, the logarithmic average of these virtual test points, and the re-assessed safe life limit SLL = 1.0 FLEI for the web of a fuselage bulkhead from CBs, using an experience-based SF = 2.8 [12]. The data points shown were taken from full-scale fatigue tests (FT55, ST16) and from service cracks found in retired CBs (CB1, 8, 9 & 12))

---

[2] That the average FLEI = 1.0 in the example provided in Fig. 6 was purely fortuitous.

6) From the pooled estimates of virtual test points calculate the logarithmic average of the FLEI values at the critical crack sizes a_{RS} (or $a_{cr} = a_{RS}$). Divide this average by the scatter factor SF to obtain the new (or re-assessed) safe life limits for each area. Fig. **6** gives an example for the web of a fuselage bulkhead in a F/A-18 Hornet aircraft [12]. The minimum web thickness, 4.32 mm, was conservatively chosen to be the critical crack depth $a_{cr} = a_{RS}$. Note that the re-assessed safe life limit lies below all the virtual test points, i.e. it is conservative.

Each step of this framework requires careful and expert consideration, but the final step, exemplified by Fig. **6**, is considered to provide a reasonable and reliably conservative estimate of the SLL. The examples in subsections **7.2** and **7.3** have been chosen to show some of the many issues that may be involved. More are given in Ref. [12].

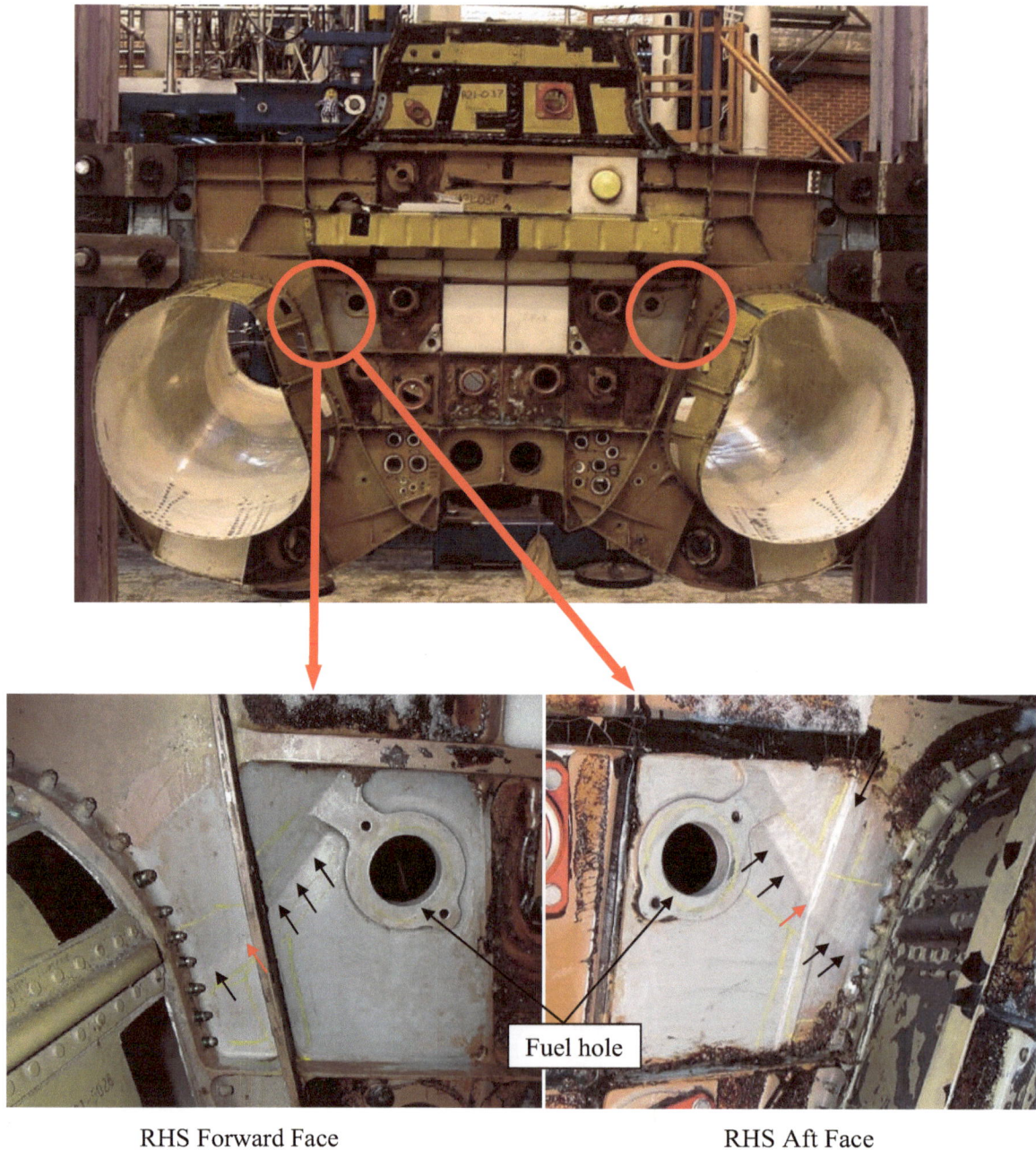

RHS Forward Face   RHS Aft Face

Figure 8: A centre barrel removed from an F/A-18A/B fuselage, showing the forward bulkhead, the two web taper areas (circled) and details of the right hand side (RHS) web taper area. The small arrows in the lower photographs indicate the fatigue-critical location, with the red arrows pointing to the most critical position along the taper.

7. EXAMPLES OF LEAD CRACK FCG ESTIMATES

7.1. Introduction

The examples of lead crack FCG estimates presented in this section refer to the same location and area, the fatigue- and fracture-critical web tapers of one of the three bulkheads in the CB of the Boeing F/A-18A/B fuselage. The CB bulkheads are machined from thick AA7050-T7451 aluminium alloy plates.

Fig. **8** shows an ex-service CB and indicates the two web taper areas and one of the critical locations in the forward bulkhead. This CB and several others were subjected to post-service full-scale fatigue testing. They provided many of the data used for the lead crack fatigue lifing summarised by Fig. **7**.

The post-service fatigue testing programme is being continued with additional CB bulkheads.

7.2. EPS, a_{RS} and Lead Crack FCG Estimates

If there is no information on initial discontinuities, e.g. when a fatigue crack has been blended out, then EPS estimates may be possible by back-extrapolating FCG data for cracks in the same area or location. This scenario applied [12] to a particular CB location in which cracks occurred during a full-scale fatigue test that was part of an international programme. It was also necessary to estimate $a_{cr} \equiv a_{RS}$.

The full-scale fatigue test was done on a fuselage with a very short service life. The bulkhead was in the as-manufactured condition. The fuselage was loaded with a service representative fatigue load history. During the test the RHS web taper area of the bulkhead in question was modified in two ways. Firstly, this area was shot peened as part of a series of modifications at 11,375 SFH. Then when cracks were found at 12,656 SFH they were blended out to enable the test to continue. The test ended successfully at the target life of 16,800 SFH.

Post-test NDI detected cracks in the RHS web taper area and near the blended-out locations. These cracks were broken open and the two deepest ones were selected for QF measurements of FCG. Fig. **9** gives the results and also a lead crack FCG estimate.

The lead crack FCG estimate was made as follows:

1)   From Fig. **9** it is seen that the QF data showed approximately exponential FCG for both cracks (C1 and C2). This enabled straightforward back-extrapolation to SFH = 0. Both extrapolations gave an EPS of about 0.01mm, which was therefore assumed for the blended-out cracks.

2)   The blended-out cracks were assumed to have just penetrated the full thickness of the web taper (4.32mm at the crack locations) when they were removed. This value was taken as a conservative critical crack size, a_{RS}.

3)   Since the QF data suggested approximately exponential FCG for cracks C1 and C2, a lead crack FCG estimate was made by connecting the assumed EPS and a point given by a_{RS} and the SFH at the blending-out time: see the red line in Fig. **9**.

Fig. **9** also shows that the lead crack FCG estimate for the blended-out cracks is conservative with respect to the known behaviour of the two nearby cracks. This estimate is therefore consistent with the earlier detection* of the blended-out cracks.

Finally, the lead crack FCG estimate was used to estimate a virtual test point to include in steps (4) − (6) of the lead crack fatigue lifing framework.

*N.B: it is important to distinguish between the *detection* and *occurrence* of cracks. Although the nearby cracks were found *at the end of* testing, the QF data in Fig. **9** suggest that these cracks grew almost form the start of the full-scale fatigue test. This would also be the case for the lead cracks, which by definition grew faster and were therefore detected *during* testing.

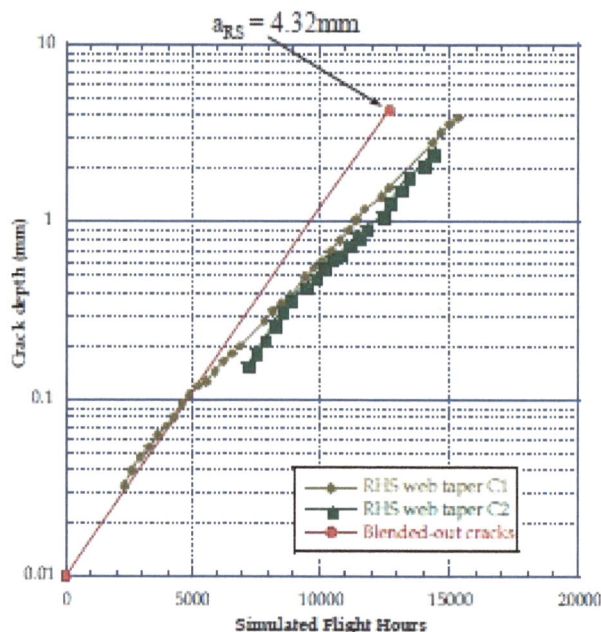

Figure 9: FCG data and a predicted lead crack FCG plot for the RHS web taper area of an F/A-18A/B production CB [12].

7.3. Lead Crack FCG Estimates from Multiple Cracks

This example comes from an ex-service CB that was subsequently tested without modification of the web taper areas. At the end of the test cracks were detected in both web tapers. However, this example is not as straightforward as the one in subsection **7.2**. This is because the fatigue load history was designed to test other structural areas and was not fully representative for the web tapers [12]. Nevertheless, the test results were useful for lead crack FCG estimates, as will be shown.

The cracked web taper areas were broken open and found to contain multiple fatigue cracks, some of which had coalesced in the RHS web taper. Figs. **10** and **11** show macroscopic views of one of the fracture surfaces from each web taper. Several cracks were selected for QF measurements of FCG. The reason to do this was that, although the local fatigue load history was not fully representative, it provided easily measured crack front markers [12, 35] that would enable determining whether FCG in the web taper areas was approximately exponential.

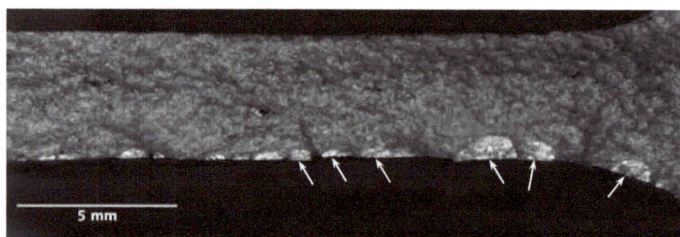

Figure 10: A fracture surface of the broken-open fatigue cracked area in the LHS web taper from an ex-service CB that was tested until cracking was detected. Some of the exposed fatigue cracks are pointed out.

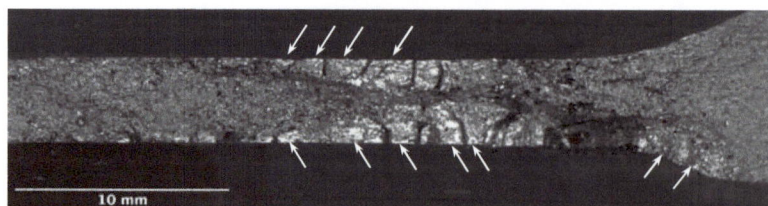

Figure 11: A fracture surface of the broken-open fatigue cracked area in the RHS web taper from an ex-service CB that was tested until cracking was detected. Some of the exposed fatigue cracks are pointed out.

Fig. **12** presents the QF results. There are three important points to note:

1) The FCG measurements began at the crack depths reached in service. In other words, a full FCG plot would require the addition of crack growth from initial discontinuities up to the crack depths when testing commenced. However, a full FCG plot is not necessary at this stage of the estimation procedure.

2) Most of the cracks showed approximately exponential FCG, although some deviations occurred beyond crack depths ≥ 0.1mm, notably for cracks in the RHS web taper. These deviations were caused mainly by load shedding to an adjacent bulkhead, but were also due to crack geometry changes.

3) The lead cracks were CB12 LHS C1 and CB12 RHS C1. Thus in the light of the results shown in Fig. **12** it was considered reasonable to assume exponential FCG for the in-service cracks in the web taper areas. The following additional steps were taken to obtain the lead crack FCG estimates for these cracks:

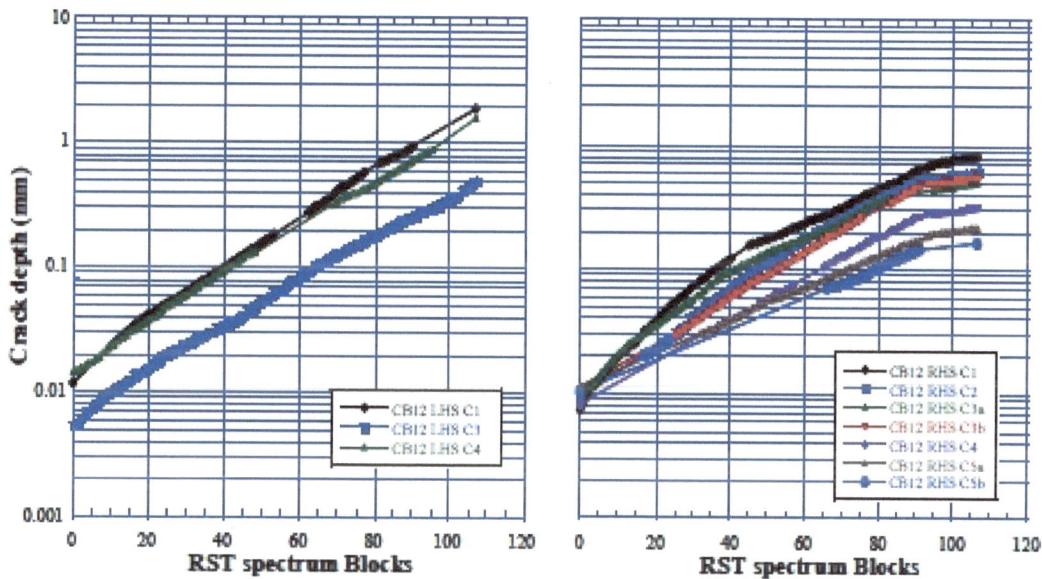

Figure 12: FCG test data for selected cracks from the web taper areas of an F/A-18A/B ex-service CB. The CB was tested with an RST spectrum developed principally to test other structural areas [12].

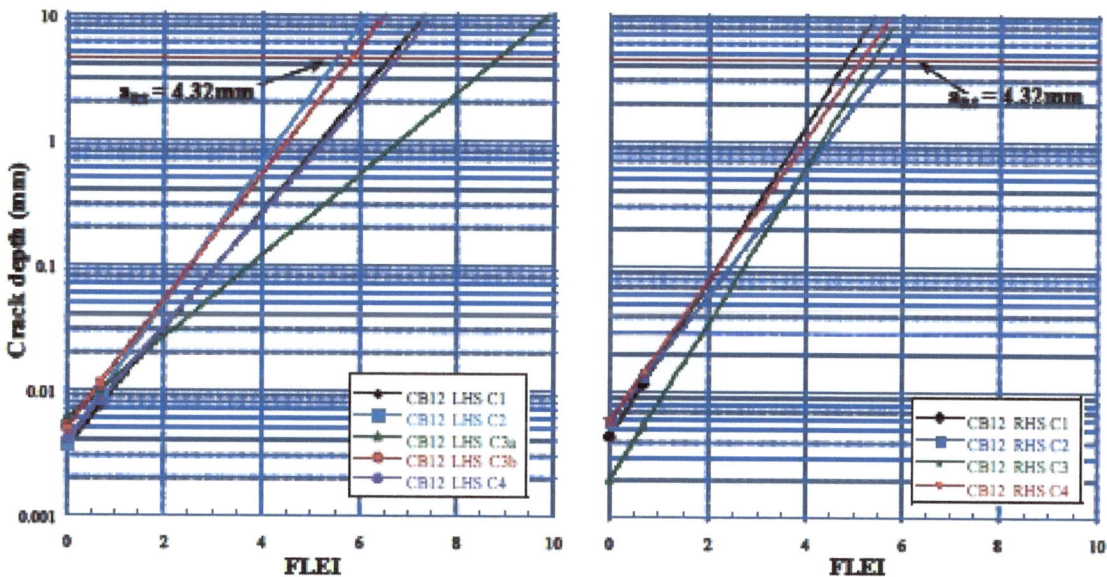

Figure 13: In-service FCG data points and conservatively-predicted FCG plots for the cracks whose test data are shown in Fig. **12**.

4) Determination of the EPS values for the initial discontinuities of the selected cracks. In this case the depths of the initiating discontinuities were used.

5) Connection of the EPS values to the crack depths reached in service, using a log crack depth *versus* flight hours plot, and extrapolation to the critical crack size in order to determine which cracks would give the shortest lives. These would then be the lead cracks.

Fig. **13** gives the results of steps (4) and (5). Note that the lead cracks were CB12 LHS C1 and CB12 RHS C1, as in Fig. **12**, but that this was not at first the case for CB12 RHS C1. A similar result of one crack overtaking another was described with reference to Fig. **2** in subsection **3.2**.

Finally, the CB12 LHS C1 and CB12 RHS C1 lead crack FCG estimates were used to estimate virtual test points for inclusion in steps (4) − (6) of the lead crack fatigue lifing framework described in subsection **6.2**.

8. CONCLUSIONS

This chapter has presented an innovative and practical approach to fatigue life assessment of high performance metallic airframe structures. This approach has been designated the lead crack fatigue lifing framework.

This framework is based on many years of detailed inspection and analysis of fatigue cracks in airframe materials and structures, ranging from coupon to full-scale fatigue tests, and also including components removed from service and subsequently tested.

The framework is now an important additional method of determining fatigue lives for aircraft in the Royal Australian Air Force fleet.

9. REFERENCES

[1] Military Specification Airplane Damage Tolerance Requirements, MIL-A-83444 (USAF), 1974.

[2] Komorowski JP, Bellinger NC, Liao M, Fillion A, "Application of the holistic structural integrity process to Canadian Forces challenges", in ASIP 2007, USAF Aircraft Structural Integrity Program 2007, Palm Springs, California, USA, Dec. 4-6, 2007.

[3] Barter S, Molent L, Goldsmith N, Jones R, "An experimental evaluation of fatigue crack growth", Eng. Fail. Anal. 2005; 12: 99-128.

[4] L. Molent L, Barter SA, "A comparison of crack growth behaviour in several full-scale airframe fatigue tests", Int. J. Fatigue 2007; 29:1090-1099.

[5] Jones R, Molent L, Pitt S, "Understanding crack growth in fuselage lap joints", Theor. Appl. Fract. Mech. 2008; 49: 38-50.

[6] Molent L, Singh R, Woolsey J, "A method for evaluation of in-service fatigue cracks", Eng. Fail. Anal. 2005; 12: 13-24.

[7] Underhill PR, DuQuesnay DL, "The effect of dynamic loading on the fatigue scatter factor for Al 7050", Int. J. Fatigue 2008; 30: 614-622.

[8] Liao M, Benak T, Renaud G, *et al.*, "Development of short/small crack model for airframe material: 7050 aluminum alloys", in 11th Joint NASA/FAA/DOD Conference on Aging Aircraft, Phoenix, Arizona, USA, Apr. 21-24, 2008.

[9] Mohanty JR, Verma BB, Ray PK, "Prediction of fatigue life with interspersed mode-I and mixed-mode (I and II) overloads by an exponential model: extensions and improvements", Eng. Fract. Mech. 2009; 76: 454-468.

[10] Molent L, Barter SA, Wanhill RJH, "The lead crack fatigue lifing framework", DSTO-RR-0353, Defence Science and Technology Organisation, Melbourne, Australia, 2010.

[11] Molent L, "Fatigue crack growth from flaws in combat aircraft", Int. J. Fatigue 2010; 32: 639-649.

[12] Barter S, Molent L, Robinson L, "Using in-service F/A-18A/B aircraft fatigue cracking as disclosed by teardown to refine fleet life limits", in ASIP 2009, USAF Aircraft Structural Integrity Program 2009, Jacksonville, Florida, USA, Nov. 30 - Dec. 2, 2009.

[13] Molent L, Sun Q, Green AJ, "Characterisation of equivalent initial flaw sizes in 7050 aluminium alloy", Fatigue Fract. Eng. Mater. Struct. 2006; 29: 916-937.

[14] Barter SA, "Fatigue crack growth in 7050T7451 aluminium alloy thick section plate with a glass bead peened surface simulating some regions of the F/A-18 structure", DSTO-TR-1477, Defence Science and Technology Organisation, Melbourne, Australia,   2003.

[15] Walker K, Weller S, Walker J, "F-111 wing pivot upper plate critical features fatigue assessment - pre wing optimisation modification", DSTO-TR-1682, Defence Science and Technology Organisation, Melbourne, Australia, 2005.

[16] Boykett R, Walker K, Molent L, "Sole operator support for the RAAF F-111 fleet", in 11th Joint NASA/FAA/DOD Conference on Aging Aircraft, Phoenix, Arizona, USA, Apr. 21-24, 2008.

[17] Wanhill RJH, Barter SA, "Fatigue of β processed and β heat-treated titanium alloys: A contribution to the DSTO – NLR joint programme of Damage Tolerance and Durability assessment of beta annealed Ti-6Al-4V plate", NLR-TP-2009-036, National Aerospace Laboratory NLR, Amsterdam, the Netherlands, 2009.

[18] Zhuang W, Barter S, Molent L, "Flight-by- flight fatigue crack growth assessment", Int. J. Fatigue 2007; 29: 1647-1657.

[19] Murakami Y, Miller KJ, "What is fatigue damage? A viewpoint from the observation of low cycle fatigue process", Int. J. Fatigue 2005; 27: 991-1005.

[20] Payne J, Welsh G, Christ RJ, Jr., Nardiello J, Papazian JM, "Observations of fatigue crack initiation in 7075-T651", Int. J. Fatigue 2010; 32: 247-255.

[21] Wanhill RJH, "Fractography of fatigue crack propagation in 2024-T3 and 7075-T6 aluminum alloys in air and vacuum", Metall. Trans. A 1975; 6A: 1587-1596.

[22] Frost NE, Dugdale DS, "The propagation of fatigue cracks in test specimens", J. Mech. Phys. Solids 1958; 6: 92-110.

[23] McClung RC, Chan KS, Hudak SJ, Jr., Davidson DL, "Behavior of small fatigue cracks", in ASM Handbook Volume 19 Fatigue and Fracture, S.R. Lampman *et al.*, Eds., ASM International, Materials Park, Ohio, USA, 1996, pp. 153-158.

[24] Wanhill RJH, Hattenberg T, "Fractography-based estimation of fatigue crack "initiation" and growth lives in aircraft components", NLR-TP-2006-184, National Aerospace Laboratory NLR, Amsterdam, the Netherlands, 2006.

[25] Hoeppner DW, Krupp WE, "Prediction of component life by application of fatigue crack growth knowledge", Eng. Fract. Mech. 1974; 6: 47-70.

[26] Paris PC, Erdogan F, "A critical analysis of crack propagation laws", J. Basic Eng., Trans. ASME, Series D 1963; 85: 528-534.

[27] Various authors in Corrosion Fatigue: Chemistry, Mechanics and Microstructure, O.F. Devereux, A.J. McEvily, and R.W. Staehle, Eds., National Association of Corrosion Engineers, Houston, Texas, USA, 1972.

[28] Various authors in Corrosion Fatigue of Aircraft Materials, AGARD Report No. 659, Advisory Group for Aerospace Research and Development, Neuilly-sur-Seine, France, 1977.

[29] Wanhill RJH, "Aircraft corrosion and fatigue damage assessment (USAF ASIP publication)", NLR TP 95656 U, National Aerospace Laboratory NLR, Amsterdam, the Netherlands, 1995.

[30] Wanhill RJH, Jacobs FA, Schijve J, "Environmental fatigue under gust spectrum loading for sheet and forging aircraft materials", in Fatigue Testing and Design, R.G. Bathgate, Ed., The Society of Environmental Engineers, Buntingford, UK 1976; 1: 8.1-8.33.

[31] Schijve J, Jacobs FA, Tromp PJ, "Environmental effects on crack growth in flight-simulation tests on 2024-T3 and 7075-T6 material", NLR TR 76104 U, National Aerospace Laboratory NLR, Amsterdam, the Netherlands, 1976.

[32] Wanhil RJH, "Flight simulation environmental fatigue crack propagation in 2024-T3 and 7475-T761 aluminium", in ICAS Proceedings 1980: 12th Congress of the International Council of the Aeronautical Sciences, J. Singer and R.W. Staufenbiel, Eds., American Institute of Aeronautics and Astronautics, Inc., New York, New York, USA, 1980, pp. 645-651.

[33] Wanhill RJH, Jacobs FA, Schra L, "The effect of environment on fatigue crack propagation under gust spectrum loading in aluminium alloy sheet, and the significance for realistic testing", in The Influence of Environment on Fatigue, The Institution of Mechanical Engineers, London, UK, 1977, pp. 101-109.

[34] Murakami Y, Endo M, "Effects of hardness and crack geometries on ΔK_{th} of small cracks emanating from small defects", in The Behaviour of Short Fatigue Cracks, K.J. Miller and E.R. de los Rios, Eds., Mechanical Engineering Publications Ltd, London, UK 1986; pp. 275-293.

[35] Barter SA, Molent L, Wanhil RJHl, "Marker loads for quantitative fractography of fatigue cracks in aerospace alloys", in ICAF 2009: Bridging the Gap between Theory and Operational Practice, M.J. Bos, Ed., Springer Netherlands, Dordrecht, the Netherlands, 2009; pp. 15-54.

CHAPTER 2

A Generic Design Procedure for the Repair of Acoustically Damaged Panels

R.J. Callinan, C.H. Wang[*], S.C. Galea and S. Sanderson

*DSTO, Melbourne, Australia and *Sir Lawrence Wackett Aerospace Centre, School of Aerospace, Mechanical and Manufacturing Engineering, RMIT University, Victoria, Australia.*

Abstract: Acoustic fatigue is the major damage phenomenon induced by the high frequency lateral vibration of structural panels, such as that of an aircraft skin, under time varying pressure waves generated by engine and/or aerodynamic effects. For instance, acoustically-induced cracks have been discovered in the lower external surface of the nacelle skin and aft fuselage of the F/A-18 aircraft. In the case of the inlet nacelle overall sound pressure levels of the order of 172 dB have been recorded. Attempts to repair these cracks by applying standard bonded repairs developed for in-plane loads were made. However the cracks continued to grow at a similar rate as before the application of repairs. While the repair of cracked aircraft structures subjected to in-plane loads using bonded repairs has resulted in considerable aircraft life extension and hence cost savings, the use of bonded patches to repair panels with acoustically-induced cracks (acoustic fatigue) is only recent. In this chapter a generic design procedure is presented for the repair of aircraft panels containing acoustically induced cracks by incorporating the constrained layer damping technique. The analytical tools described in this chapter will enable the rapid design of damped repairs using closed form solutions that account for the effects of high frequency out-of-plane vibration on crack extension. A case study is also undertaken of a design of a repair to prevent acoustic fatigue cracking on the aft fuselage of the F/A-18.

INTRODUCTION

Adhesively bonded patch repairs to cracked aircraft structures subjected to in-plane loads have resulted in considerable aircraft life extension and hence cost savings [1,2]. However, the use of bonded patches to repair panels with acoustically induced cracks (acoustic fatigue) is a recent development. In this chapter, we will present a generic design procedure for repairs of panels containing acoustically-induced cracks. Acoustic fatigue is the result of high frequency lateral vibration due to time varying pressure waves generated by engine and/or aerodynamic effects. For instance, acoustically- induced cracks have been recorded in the lower external surface of the nacelle skin of the F/A-18 aircraft, as shown in Fig. 1. In this case, overall sound pressure levels of the order of 172 dB have been reported [3,4]. Attempts were made to repair these cracks by applying standard methods of bonded repair [5] developed for in-plane loads, however, the cracks continued to grow. It was found [6] that while the boron fibre composite repairs did reduce the stress intensity at the crack-tip, the high number of cycles still led to significant crack growth of up to 2*mm* per flight hour. Investigations by the present authors revealed that the main cause for the observed ineffectiveness of standard repairs was the neglect of high frequency out-of-plane vibration [7]. Consequently a major effort was directed at developing new analysis methods to improve design and incorporate new damping technique to improve repair efficiency [8].

Since it is known that the amplitude of vibration is inversely proportional to the square root of the damping ratio, a combination of damping and stiffness may reduce the crack growth rate significantly. The application of highly damped repairs (Durability Patches) to acoustically damaged panels was proposed by Rogers *et al* [9]. Also, a highly damped repair was applied [10] to the vertical fin of the F-15 aircraft, as part of a test program. More recently, a series of repairs have been designed and tested in the Air Force Research Laboratory progressive wave tunnel [11, 12]. In addition, Boeing [13] developed a procedure for the use of a Dosimeter, and methods for optimal design of damped repairs. The Dosimeter was employed to assess the effectiveness of repairs made to B52 fuselage skin panels, an F-15 access panel and C130 flap panels [13]. The Dosimeter being a small autonomous device, carried in an aircraft, and designed to measure the thermo-acoustic environment of an aircraft skin plate.

In this chapter, a generic design procedure, based on constrained layer damping (CLD) will be presented. This will involve the selection of a damping material and the geometry of the CLD in order to significantly reduce the growth rate or even completely arrest the growth of acoustically-induced cracks. The present work makes extensive use of the closed form solutions of bridged cracks by adhesively bonded patches [2,7,8].

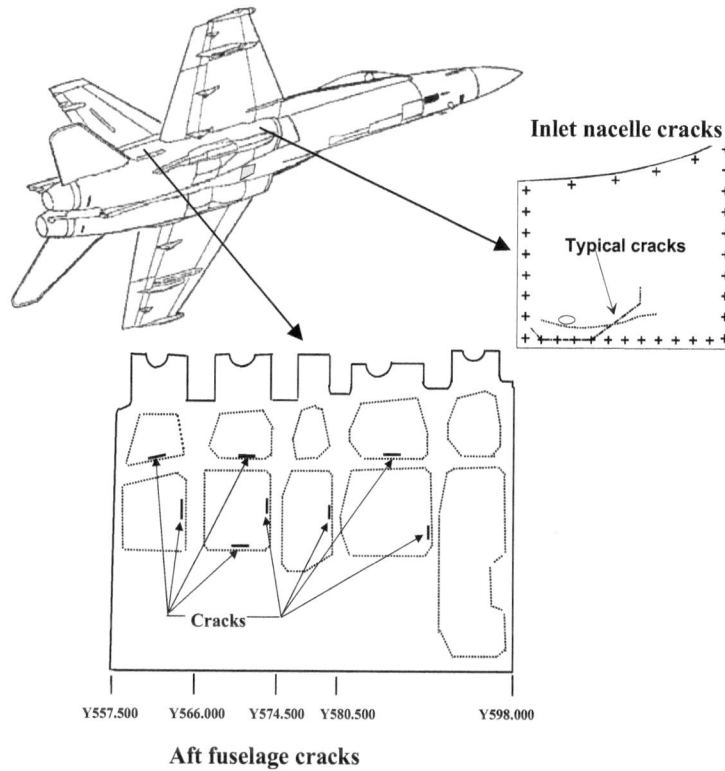

Figure 1: Location of acoustic fatigue cracking in F/A-18 fuselage panels.

OVERVIEW OF DESIGN PROCEDURE

The use of CLD containing a viscoelastic material is well known as an effective method of dissipating the energy due to flexural vibration.

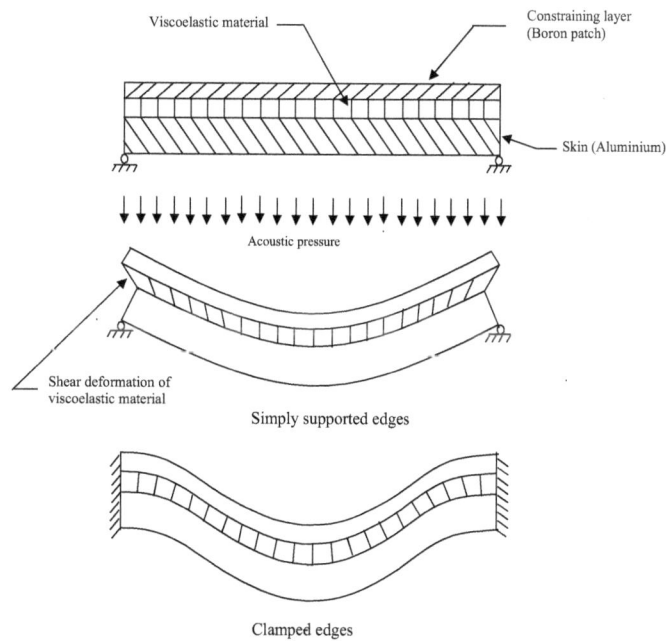

Figure 2: Shear deformation of viscoelastic layer under constrained layer damping.

As shown in Fig. **2**, CLD involves the use of a constraining layer to attach a viscoelastic material to one side of the vibrating panel. In the present context, the constraining layer is effectively the repair patch. The bending of the structure involves intense shear deformation of the viscoelastic material, which dissipates energy via friction of its long chain molecules. A measure of the effectiveness of the CLD is the loss factor, η. For a vibrating plate, the loss factor may be defined as the ratio of the energy dissipated per radian to the total strain energy. The viscoelastic material has a very high loss factor in comparison to metallic components. To achieve the maximum efficiency of energy dissipation, the constraining layer should be very stiff in order to promote shear deformation of the viscoelastic material, i.e. the use of boron/epoxy composite material.

It is known that the amplitude of vibration at a resonant frequency is inversely proportional to the square root of the loss factor. Hence the higher the loss factor the lower the amplitude of vibration and stress in the plate. Shown in Fig. **3** is a plot of shear modulus and loss factor versus temperature for a viscoelastic damping material. Three distinct states exist, namely, glassy, transition and rubbery. The maximum loss factor occurs in the transition state. It is evident that this type of viscoelastic damping materials is optimum over a limited temperature range. The energy dissipation response is clearly dependent on resonant frequency. Hence the correct material must be used for the expected environment (temperature and resonant frequency). The material response for different damping materials will be considered in Section 3.

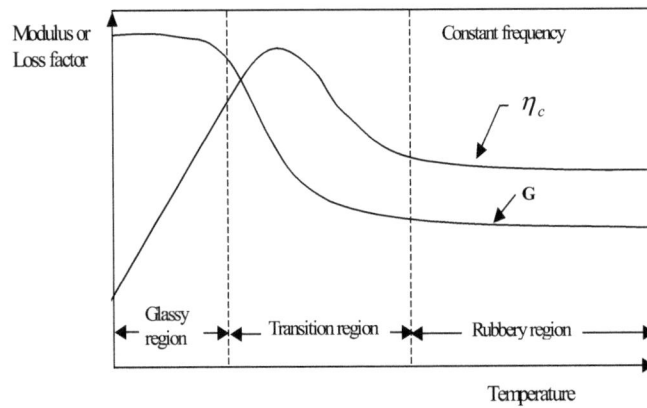

Figure 3: Plot of shear modulus (G) and loss factor (η) as a function of temperature at constant frequency for a viscoelastic material.

To simplify the analysis of damped repairs, an analytical solution to a three-layered plate will be considered in order to obtain the loss factor and the resonant frequency. This analysis will be implemented using a symbolic solver, Mathematica^{TM}.

Consider a simply-supported rectangular plate, as shown in Fig. **4**, featuring a single viscoelastic layer and a constraining layer.

The moduli of elasticity of the plate and the constraining layer are denoted as E_1 and E_3 respectively, and the damping layer is characterized by its shear modulus G_2. The flexural stiffness of layers 1 and 3 are E_1I_1 and E_3I_3, where I_1 and I_3 are the second moment of area of the cross-sections of layer 1 and 3 respectively.

Let the plate lie in the x-y plane, then the governing equation of the displacement, w, in the z direction is given by Mead and Markus [14]:

$$EI_t \frac{\partial^6 w}{\partial x^6} - g^* EI_t (1+Y) \frac{\partial^4 w}{\partial x^4} + m_t \left[\frac{\partial^4 w}{\partial x^2 \partial t^2} - g^* \frac{\partial^2 w}{\partial t^2} \right] = \frac{\partial^2 p(x,t)}{\partial x^2} - g^* p(x,t) \qquad (1)$$

where $p(x,t)$ is space-time-dependent acoustic pressure, and m_t is the total mass per unit length of the beam. The total bending stiffness is

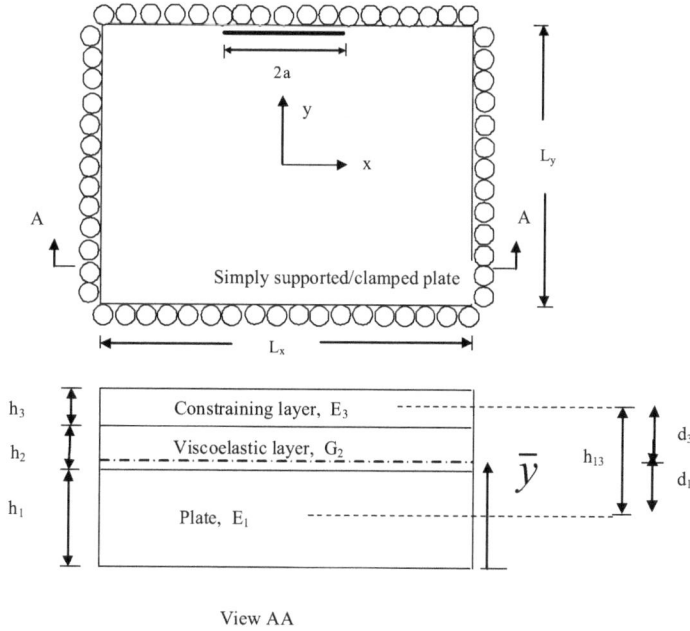

Figure 4: Geometry for constrained layer damping of a simply supported plate.

$$EI_t = E_1 I_1 + E_3 I_3 \tag{2}$$

The shear parameter g^* is defined by:

$$g^* = \frac{G_2}{h_2}\left(\frac{1}{E_1 h_1} + \frac{1}{E_3 h_3}\right) \tag{3}$$

with the geometric factor Y [p389, 14] being

$$Y = \frac{h_{13}^2}{D_t(1-v^2)}\left(\frac{E_1 h_1 E_3 h_3}{E_1 h_1 + E_3 h_3}\right) \tag{4}$$

where D_t is the total flexural rigidity and will be defined later, while v is the Poisons ratio. The solution of this equation leads to the explicit relationship for the loss factor for a three-layer plate given by [p389, 14],

$$\eta_S = \frac{g_m \eta_c Y}{1 + g_m(2+Y) + g_m^2(1+\eta_c^2)(1+Y)} \tag{5}$$

where g_m is the geometric shear parameter and η_c is the loss factor for the viscoelastic material. To accurately predict the modal loss factor, particularly at low temperatures when the loss factor of the damping material is low, in the following we present an extension of the Mead's solution [15] to include the viscoeleastic loss of the skin and the constraining layer.

Denoting the loss factors of the skin panel, the constrained layer, and the constraining layer as η_1, η_2, and η_3, respectively. The elastic properties of these three layers are denoted as E_1, G_2, E_3, which are given by

$$E_1 = E_1'(1 + i\eta_1) \tag{6}$$

$$G_2 = G_2'(1 + i\eta_2) \tag{7}$$

$$E_3 = E_3'(1 + i\eta_3) \tag{8}$$

with E_1', E_3', and G_2' denoting real parts of the complex Young's moduli of the plate, of the constraining layer, and the shear modulus of the viscoelastic layer, respectively. The shear parameter is thus

$$g = g'(1+i\eta_S)$$ (9)

where g' is the real part of the complex shear parameter and,

$$\eta_S = \frac{E_1 h_1 (\eta_2 - \eta_3)(1+\eta_1^2) + E_3 h_3 (\eta_2 - \eta_1)(1+\eta_3^2)}{E_1 h_1 (1+\eta_2 \eta_3)(1+\eta_1^2) + E_3 h_3 (1+\eta_1 \eta_2)(1+\eta_3^2)}$$ (10)

It is readily seen that the above solution gives $\eta_S = \eta_2$ when $\eta_1 = \eta_3 = 0$. (11)

thus recovering the Mead solution [15].

The total bending stiffness of the repaired structure is

$$EI_t = E_1 I_1 + E_3 I_3 = D_t (1+i\eta_t)$$ (12)

with η_t being defined as the non-core loss factor:

$$\eta_t = \frac{E_1' I_1 \eta_1 + E_3' I_3 \eta_3}{E_1' I_1 + E_3' I_3}$$ (13)

The *m*th complex frequency, given by equation (9.52) in ref [14] can be written as

$$\omega_m^2 (1+i\eta_{DC,m}) = \frac{EI_{tr}(1+i\eta_t)k_f^2}{m_t} \left[1 + Y \frac{g_m(1+i\eta_S)}{1+g_m(1+i\eta_S)}\right]$$ (14)

where

$$g_m = g'/k_f^2 = \frac{G_2'}{k_f^2 h_2}\left(\frac{1}{E_1 h_1} + \frac{1}{E_3 h_3}\right)(1-v^2)$$ (15)

The modal loss factor $\eta_{DC,m}$ is the ratio between the imaginary part and the real part of the expression on the right hand side of the equation (13). The effective wave number k_f will be defined later. After some simplification, the modal loss factor becomes,

$$\eta_{DC,m} = \frac{\eta_t + \eta}{1 - \eta_t \eta}$$ (16)

with

$$\eta = \frac{g_m \eta_S Y}{1 + g_m(2+Y) + g_m^2 (1+\eta_S^2)(1+Y)}$$ (17)

which is the original solution by Mead, without considering the structural loss due to the skin panel and the constraining layer (the repair patch).

Since $\eta_t \ll 1$ for most metallic materials, the modal loss factor can be approximated by

$$\eta_{DC,m} = \eta_t + \eta$$ (18)

Hence the modal loss factor will equal the sum of non-core loss and the loss factor due to the constrained layer. A value of η_t for 7075-T6 at room temperature is available from ref [15]: $\eta_t = 2\delta = 0.0106$. Here the symbol δ is the material damping ratio. This value will not change significantly with temperature or frequency.

For built-up aircraft structures the skin damping is enhanced by frictional damping due to fastening connections, resulting in typical loss factors of the order of 3%.

As indicated the geometric shear factor can be expressed as

$$g_m = g' / k_f^2 \tag{19}$$

and for two-dimensional structures, the shear parameter is related to the wave numbers in two mutually perpendicular directions, which is defined by

$$k_f^2 = k_x^2 + k_y^2, \tag{20}$$

where the modal wave numbers in the x and y directions for the fundamental mode are given by, $k_x = \pi / L_x$ for simply supported, $k_x = 2\pi / L_x$ for clamped edge condition. Similarly, $k_y = \pi / L_y$ for simply supported, $k_y = 2\pi / L_y$ for clamped edges. Here L_x and L_y are the lengths of the plate in the x and y directions respectively.

It can be shown that for a three-layer beam the optimum loss factor is given by [p392, 14] and is modified:

$$\eta_{S,\max} = \frac{\eta_c Y}{(2+Y) + 2\sqrt{(1+Y)(1+\eta_c^2)}} + \eta_t \tag{21}$$

and an optimum value of the shear parameter exists, and is given by [p392, 14]:

$$g_{opt} = \frac{1}{\sqrt{(1+Y)(1+\eta_c^2)}} \tag{22}$$

If the optimum value of the loss factor given by equation (21) is much higher than that given by equation (17), a design change should be considered. If aerodynamic restrictions allow the repair to be thickened, increasing parameters h_3 and/or h_2 should be considered. Also the lay-up of the constraining layer should allow for a maximum value of E_3. A comparison of the shear parameter, equation (9), with the optimum value given by equation (22) will also provide some indications on how far the design can be improved. All the design parameters are given in Table **1**.

Table 1: Design parameters including dimensions, properties and parameters.

Physical dimensions (Fig. 4)	Material properties	Key parameters
h_1, h_2, h_3, h_{13} L_x, L_y & $2a$	E_1, E_3, G_2 $v_1, v_3,$ ρ_1, ρ_2 & ρ_3	$\eta_s, \eta_c, \eta_t, g_m, k_x, k_y,$ $k_f, f_o,$ K$_{rms}$, & Y

DESIGN PROCESS

The design process is illustrated in Fig. **5**. Firstly an initial guess is required for the CLD treatment for the structure.

Note that in Fig. **5** the sound pressure level (SPL) is required. This will be defined in a later section. The shear modulus, loss factor, Young's modulus, temperature, frequency and SPL, are necessary inputs. From these quantities geometric shear factor, geometric factor and loss factor are calculated. It may be useful to first examine the optimality for this geometry. If the design is not optimum, the dimensions of the damping layer can be adjusted by changing h_2 or h_3. For the first design iteration it is possible to neglect the optimum requirement. Firstly determine out-of-plane displacement and resonant frequencies. Secondly calculate the stress intensity factor due to the SPL and residual thermal stresses. Likely crack growth rates and reductions are then assessed. Iteration completes when the design becomes acceptable over the entire operating temperature range and loading regime. It may be necessary to choose an alternative damping material if results are not acceptable.

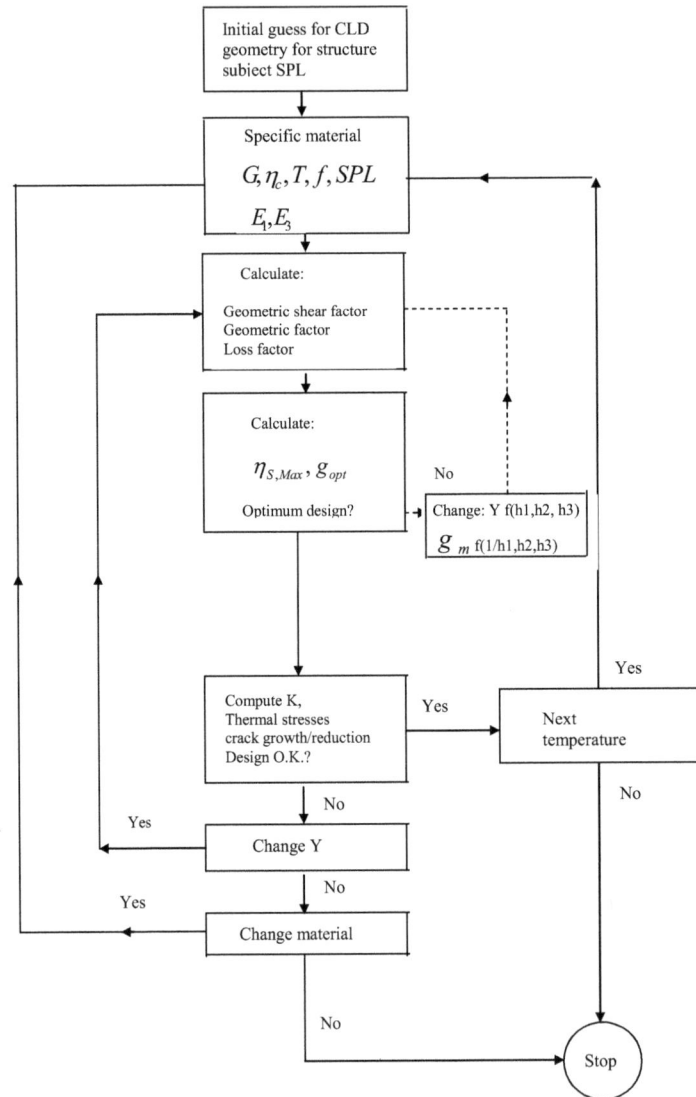

Figure 5: Flow chart for design of highly damped repairs.

A schematic of the proposed repair design is shown in Fig. **6**. For best results the edge of the repair ideally should terminate away from supporting substructures, such as frames or stiffeners, to allow for shear deformation of the viscoelastic damping layer. Otherwise the maximum possible loss factor will not be achieved. Note that both an erosion layer and sealant should be included to protect the repair from environmental degradation and moisture ingress into the adhesive bond.

Design of the repair shown in Fig.6 considered the following issues:

(a)   The constraining layer was selected to have a stiffness similar to that of the plate being repaired.

(b) The damping material was chosen such that optimal damping occurs within the expected temperature range. Two viscoelastic materials, Soundcoat Dyad 606 and 609, which attain their respective peak performance at 30°C and 50°C, were chosen in the present investigation. The values of η_2 and shear modulus G_2' can be obtained from Fig.7 or Fig.8 [16]. The thickness of the damping layer must be specified for a given application.

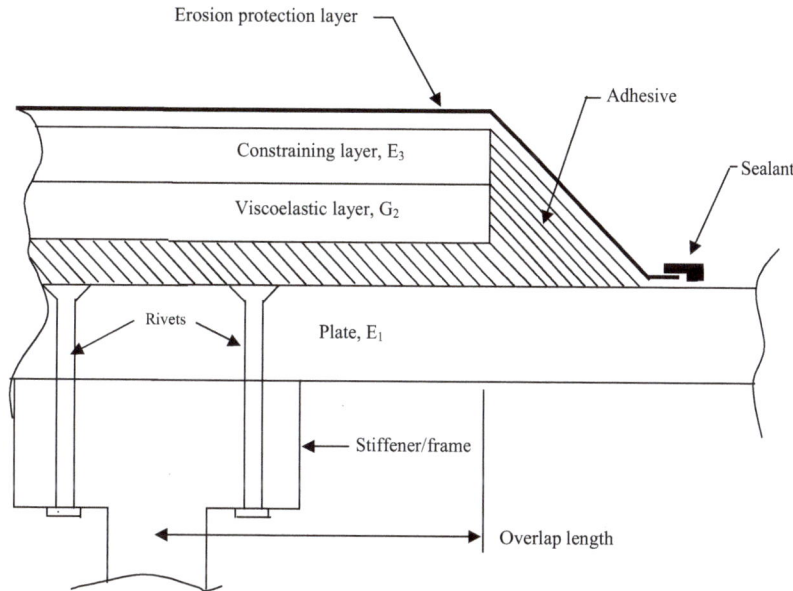

Figure 6: Edge detail necessary to maximize the shear in the viscoelastic layer.

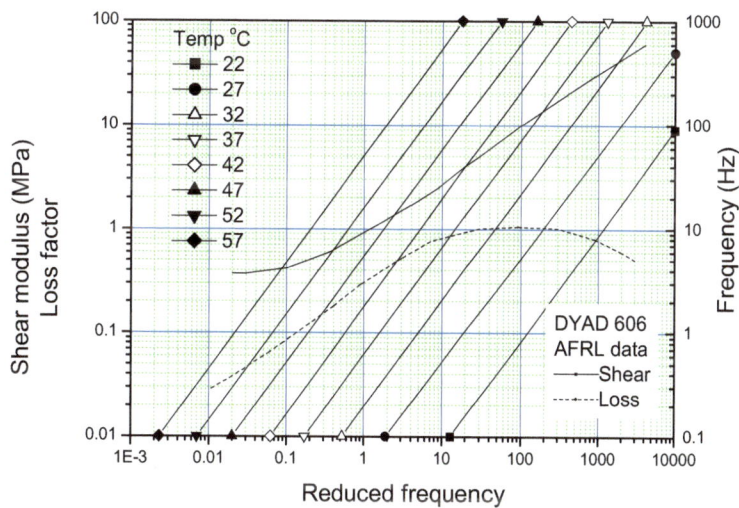

Figure 7: Dyad 606 material data, [15].

(c) The damping properties of Soundcoat Dyad 606 and 609 can be obtained from the plots shown in Fig. **7** and using the following steps, referring to Fig. **9**:

- For the specific panel resonant frequency (right hand scale) project horizontally to intersect the appropriate oblique temperature curve.

- Now intersect the shear (storage) modulus curve vertically and read off the shear (storage) modulus from the left hand scale.

- • Obtain the loss factor by intersecting the loss factor curve vertically and read off the corresponding loss factor from the left hand scale.

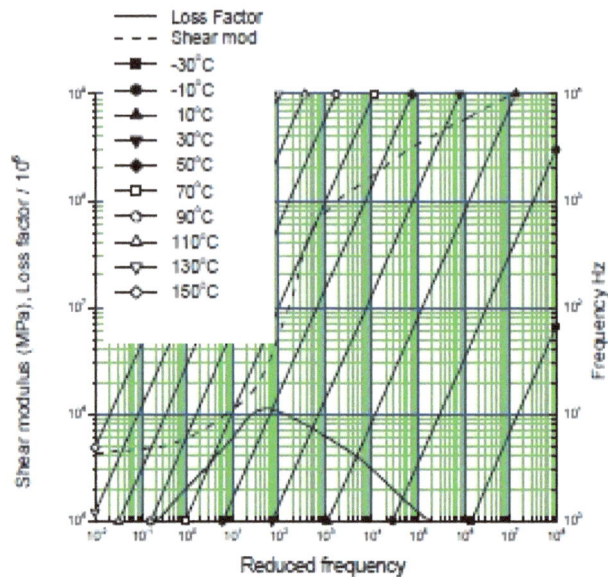

Figure 8: Dyad 609 material data. (To obtain loss factor divide shear modulus scale by 10^6).

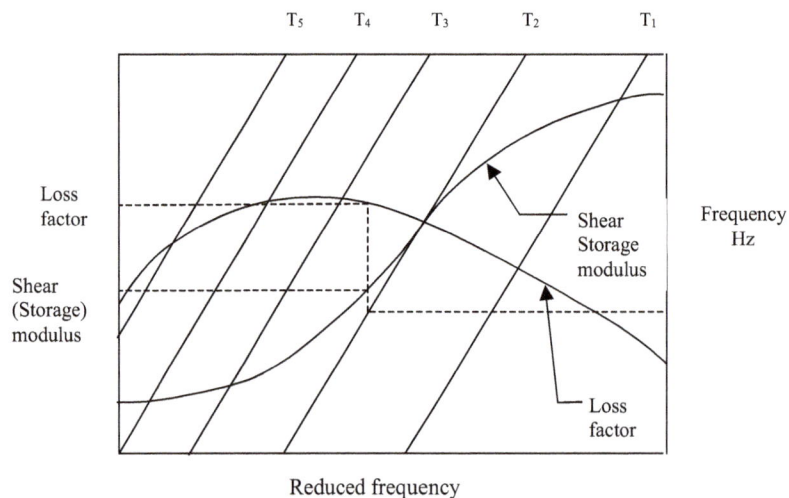

Figure 9: Procedure to determine damping properties at the required frequency and temperature. Higher values for the subscript i for T_i indicate increasingly higher temperatures.

(c)   The load to be applied was the power spectral density (PSD) of the excitation, as described in a later section.

(d)   Determine the stress intensity, stresses, strains and lateral deformation as described in a later section.

(e)   Assess the thermal residual stresses as described in a later section.

(f)   Determine crack growth rates as detailed in a later section.

Experimental evidence indicated that fatigue cracking usually occurred along the edges of the plate rather than in the centre of the plate. This suggests that, as a first approximation, clamped edge conditions are appropriate.

SOUND PRESSURE LEVELS

Sound pressure levels (SPL) measured on the external surface of an F/A-18 inlet nacelle are plotted in Fig. **10**. The spectrum level, relative to the overall sound pressure level (OASPL), was derived from in-flight one-third octave SPL measurements, [2]. This spectrum level is now used to calculate the PSD of the excitation, $S_I(f_0)$. OASPL pressure levels of 172.2dB have been measured on the inlet nacelle [2].

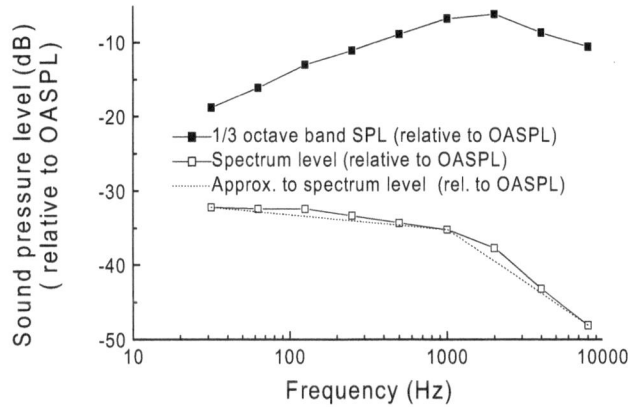

Figure 10: Spectrum and one-third octave band levels of sound pressure over the external nacelle inlet (where OASPL=172.2 dB) [2].

The relationship between the spectrum SPL and the root-mean-square (r.m.s) of the fluctuating pressure p, in (Pa), is given by [17] as

$$p_{rms} = 10^{SPL/20-4.65894} , \tag{24}$$

and the power spectral density of acoustic pressure, i.e. PSD of the excitation, at any given frequency is given by

$$S_I(f_o) = p_{rms}^2 = 10^{SPL/10-9.3979} . \tag{25}$$

The curve in Fig. **10** has been mapped by two straight lines defined by the three points listed in Table **2**. This approximate spectrum is used as the excitation PSD in the finite element analysis.

Table 2: Input power spectral density corresponding to an OASPL of 172.2 Db.

Freq. (Hz)	Three points to approximate curve (dB)	Pressure spectrum level (dB)	$S_I(f_0)$ (MPa)²/Hz
31.5	-32.2	140	4x10⁻⁸
1000	-35.2	137	2.005x10⁻⁸
8000	-48.1	124.1	1.028x10⁻⁹

COMPUTATION OF STRESS INTENSITY FACTOR

Acoustic fatigue is associated with the random loading of a structure over a wide frequency range. In the case considered here, the loading is applied to a rectangular panel shown in Fig. **4**. The complete analytical solution of the dynamic response of the panel is very complex [18]. However, in the case of acoustic fatigue, a number of assumptions can be made, as the emphasis here is to determine the crack growth driving force. Firstly, the time varying pressure can be considered to be uniform and in phase over the panel. Secondly, since the first mode contributes to 98% of the panel response [3], the contribution of all other modes will be ignored. Also the variation of power spectral density (PSD) may be assumed to be constant with frequency near the first mode frequency. Furthermore, it is assumed that the loss factor using equations (11-13) are applicable to a rectangular plate supported on all sides. The solution [11] is directly applicable to simply supported edges, however a modification may be made to apply to clamped edges. Together with the assumptions of [19], the Miles solution gives the root-mean-square (in time-space) of the stress intensity response as:

$$K^{rms} = \sqrt{\frac{\pi f_0 S_I(f_0)}{2\eta_s}} \overline{K} \tag{26}$$

where

K^{rms} is the rms stress intensity factor due to the acoustic loading,

$\overline{K}$ is the static stress intensity factor due to the application of a unit pressure on a structure,

$S_I(f_0)$ is the PSD of the excitation at f_0

f_0 is the first resonant frequency (H_z) which depends on boundary conditions, and

η_s is the loss factor

In general, aircraft skin panels are usually attached to other panels or sub-structures at their boundaries, consequently the acoustic response lies somewhere between the limiting cases of simply supported edges or fully-fixed edges. Cracking on the side suggests that clamped boundary conditions dominate.

UN-REPAIRED PLATE

Mode Shape

It is necessary to define a mode shape that satisfies the boundary conditions at the edge and centre of the plate. For clamped edge conditions a mode shape has been found from [20] to give good results. This mode shape can be represented by:

$$w(x,y) = w_0 \left[1 - 4\left(\frac{x}{L_x}\right)^2 \right]^2 \left[1 - 4\left(\frac{y}{L_y}\right)^2 \right]^2 \tag{27}$$

where w_0 is the maximum out of plane displacement at the centre of the plate. According to Miles solution, the maximum transverse displacement w_0 at the centre of a plate is,

$$w_0 = \sqrt{\frac{\pi f_0 S_I(f_0)}{2\eta_s}} \ W(L_y/L_x) \frac{L_x^4}{E_1 h_1^3} \tag{28}$$

where $W(L_y/L_x)$ is given in ref [21]; the numerical values for a clamped panel can be approximated by the expression from ref [22],

$$W = 0.0284 e^{-0.754(L_x/L_y)^{3.79}} \tag{29}$$

Combination of expressions (27-29) furnishes a closed-form expression for the displacement at any point.

RESONANT FREQUENCY

For the case of an un-repaired plate with all edges clamped, the first mode frequency is given by the following expression, assuming that the crack is small enough that it does not significantly affect the first resonant frequency,

$$f_0 = \frac{1}{2\pi} F(L_y/L_x) \sqrt{\frac{D_t}{\rho h L_x^4}} \, ,$$

(31)

where ρ is the density, L_x and L_y are the length and width of the plate respectively. The parameter $D_t = Eh^3/12(1-v^2)$ denotes the bending stiffness. The geometry factor $F(L_y/L_x)$ is given in ref [21]. For the case of clamped edges, the geometry factor can be approximated by,

$$F = e^{0.477(L_x/L_y)^{2.3}}$$

(32)

which will be useful for performing rapid parametric design and optimisation.

STRESS AND STRAIN

The deformation of an un-repaired plate under acoustic loading is of pure bending, with the bending strain along the x-direction being given by,

$$\varepsilon_b = \frac{h_1}{2} \frac{\partial^2 w}{\partial x^2}$$

(33)

Along the clamped edge $y=0$, the root-mean-square of the bending strain becomes

$$\varepsilon_b^{rms} = \sqrt{\frac{\pi f_0 S_I(f_0)}{2\eta_s}} \frac{8 W L_x^2}{E_1 h_1^2} \left[12 \left(\frac{x}{L_x} \right)^2 - 1 \right]$$

(34)

With the strain parallel to the clamped edge being zero, the rms value of the bending stress is thus

$$\sigma_b^{rms} = \sqrt{\frac{\pi f_0 S_I(f_0)}{2\eta_s}} \frac{8 W L_x^2}{(1-v_1^2) h_1^2} \left[12 \left(\frac{x}{L_x} \right)^2 - 1 \right]$$

(35)

STRESS INTENSITY FACTOR

When the crack is short such that it has negligible effect on the mode shape of the plate, the stress intensity factor for the plate alone (without any repair) subjected to uniform pressure can be determined using the superposition method. In this case, the stress intensity factor can be expressed in terms of the prospective stresses, for both simply supported and clamped plates. In particular, the bending stress intensity factor of a crack of total length $2a$ is given by,

$$K_b = \sigma_b \sqrt{\pi a \frac{1+v}{3+v}}$$

(36)

Applying the Miles solution, the root-mean-suare value of the bending stress intensity factor becomes,

$$K_b^{rms} = \sigma_b^{rms} \sqrt{\pi a \frac{1+v}{3+v}} \, .$$

(37)

with σ_b^{rms} being given by equation (35).

VALIDATION OF CLOSED FORM SOLUTION FOR UN-REPAIRED PANEL

Validation of the closed form solution has been carried out using both experimental testing and F.E. analysis. The latter involved the use of the random vibration capability of the NASTRAN finite element program. The experimental data used to validate the closed form solution was from flight trials [24].

For the panel shown in Fig. **11** consider a traverse line from point A to point B. A comparison between strains from the closed form solution, the FE analysis, and the flight trial measurements is presented in Fig. **12**. The overall sound pressure level, OASPL, is determined by matching the closed form solution and the experimental strain at the centre of the plate.

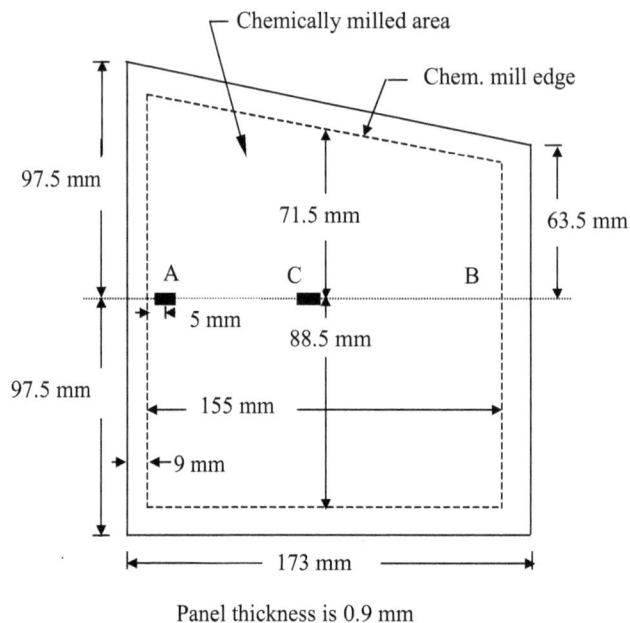

Figure 11: Dimensions of panel and of chemical milled area.

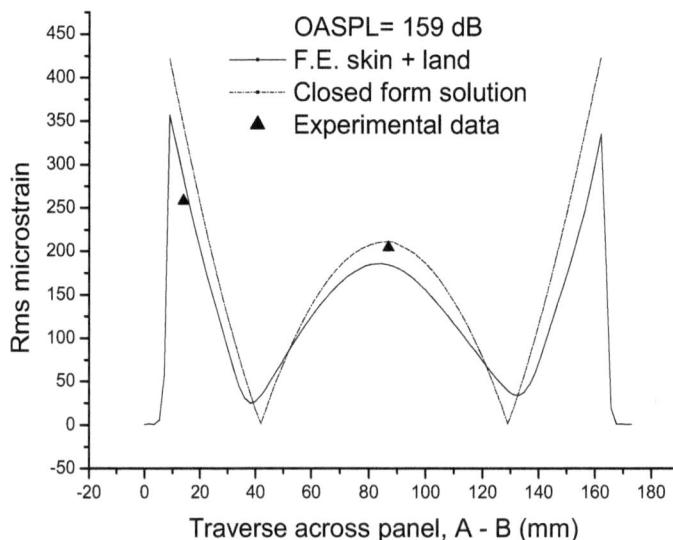

Figure 12: Rms microstrain for un-patched panel, versus distance across the panel, A to B.

The F.E. results determined at the calculated OASPL agree well with the closed form solution, and the experimental data. At the edge of the plate the closed form solution slightly overestimated the flight trial data point, while the F.E. slightly underestimated the closed form solution. Overall agreements between the closed form expression, FE analysis, and flight test data are satisfactory. As a result, the actual overall sound pressure level (OASPL) was determined to be 159 dB. This will form the basis for the design of the repair.

REPAIRED PLATE

Mode Shape and Frequency

The same mode shape for the un-repaired plate is used. As for the case of an un-repaired plate with all edges clamped, it is assumed that the crack is small enough that it does not significantly affect the first resonant frequency. When the stiffness of the visco-elastic layer is negligible, i.e., zero, the first mode frequency is,

$$f_o = \frac{1}{2\pi} F(L_y / L_x) \sqrt{\frac{D_t}{m_t L_x^4}}$$ (38)

where $m_t = \rho_1 h_1 + \rho_2 h_2 + \rho_3 h_3$ is the total mass per unit length and the bending stiffness is given by:

$$D_t = \frac{E_1 h_1^3}{12(1-v_1^2)} + \frac{E_3 h_3^3}{12(1-v_3^2)}$$ (19)

and $F(L_y / L_y)$ is given by expression (31) for clamped plate.

The correct frequency for a non-zero or finite stiffness core can be determined by the use of a non-dimensional frequency ratio,

$$\Omega = \sqrt{1 + Y \frac{g_m \left(1 + g_m (1 + \eta_S^2)\right)}{1 + 2g_m + g_m^2 (1 + \eta_S^2)}}$$ (20)

The resonant frequency for a finite-stiffness core is thus

$$f_0 = \frac{1}{2\pi} F(L_y / L_x) \sqrt{\frac{D_t}{m_t L_x^4}} \Omega$$ (41)

STRAINS AND STRESSES IN REPAIRED PLATE

Following the above derivations, the root-mean-square (RMS) out-of-plane displacement w_0 at the centre of a repaired plate is given by

$$w_0^{rms} = \sqrt{\frac{\pi f_0 S_I (f_0)}{2\eta_s}} \frac{W L_x^4}{E_1 h_1^3 + E_3 h_3^2} \frac{1}{\Omega^2}$$ (32)

Using the same mode shape given by equation (27), RMS of the out-of-plane displacement $w(x,y)$ can be expressed as,

$$w^{rms}(x,y) = w_0^{rms} \left[1 - 4\left(\frac{x}{L_x}\right)^2 \right]^2 \left[1 - 4\left(\frac{y}{L_y}\right)^2 \right]^2$$ (43)

The bending strain along $y=0$ can be derived in a similar way as for the un-repaired plate,

$$\varepsilon_b^{rms} = -\frac{\partial^2 w^{rms}}{\partial x^2} \frac{h_1}{2}$$ (44)

The r.m.s membrane strain in the plate is given by Mead and Markus, [13], viz.,

$$\varepsilon_m^{rms} = -\frac{1}{gdE_1h_1}\left[m_t\omega_0^2 w^{rms} - D_t\frac{\partial^4 w^{rms}}{\partial x^4} + gD_tY\frac{\partial^2 w^{rms}}{\partial x^2} \right] \tag{45}$$

where $d = h_2 + (h_1 + h_3)/2$, referring Fig.4.

Since the strain parallel to the clamped edge is zero, the rms stresses are

$$\sigma_m^{rms} = \frac{E_1\varepsilon_m^{rms}}{1-v_1^2}, \tag{46a}$$

$$\sigma_b^{rms} = \frac{E_1\varepsilon_b^{rms}}{1-v_1^2} \tag{46b}$$

STRESS INTENSITY FACTOR OF REPAIRED PLATE SUBJECTED TO ACOUSTIC PRESSURE

The most common type of repair applied to aircraft skins are one sided as illustrated in Fig. **13**, since, in general, access is only available to the outside of the skin. Analytical solutions for one-sided repairs, known as the Wang-Rose model are available for both in-plane [25,23] and out-of-plane loading [26]. A significant feature of bonded repairs is that the variation of stress intensity with crack length does not increase indefinitely with crack length [26,27]. Instead, the stress intensity approaches an asymptotic value. The thickness of the adhesive is included in the analysis to represent the damping material. The computation of stress intensity was based on the work carried out in ref [7].

Consider the damped plate shown in Fig. **13**. Depicted in Fig. **13a** are the components of the damped plate subjected to a lateral loading resulting in a bending moment. In the absence of a crack, the bending stress which varies linearly through the plate thickness is shown in Fig. **13a**. The Wang-Rose crack bridging theory involves firstly replacing a crack in the damped repair by springs along the face and secondly use of the lap joint analysis to determine the load transfer. This results in the following equations for the membrane stress intensity K_m and the bending stress intensity K_b, assuming the membrane and bending stresses are constant over the crack length,

$$K_m^{rms} = f_{mm}\ \sigma_m^{rms} + f_{mb}\ \sigma_b^{rms} \tag{47a}$$

$$K_b^{rms} = f_{bm}\ \sigma_m^{rms} + f_{bb}\ \sigma_b^{rms} \tag{47b}$$

where f_{mm}, f_{mb}, f_{bm} and f_{bb} are the normalised stress intensity factors. Plots of the normalised stress intensity factors are shown in Figs. **14** and **16** for temperatures of -67 $^{\circ}$C, 30 $^{\circ}$C and 120°C. Fig. **17** shows the variation of the normalised stress intensity factor with temperature.

In the present case, as the crack length increases, the prospective membrane and bending stresses vary along the crack length. In this case, a full crack bridging analysis is required to account for the effect of stress gradient. For an edge crack lying between ($-a$, $L_y/2$) and (a, $L_y/2$), the prospective membrane stress σ_m and the prospective bending stress σ_b both vary with the coordinate. The crack-bridging problem, viz, the integral equations [23], can be solved numerically using a Galerkin method: expand the unknown functions in terms of Chebyshev polynomials and then determine the coefficients numerically. Because the prospective stresses vary with the position along the crack path, the matrix equations become [22]

$$\hat{A}_{ij}\hat{f}_j + \hat{B}_{ij}\hat{g}_j = \frac{1}{E_s}\int_{-1}^{1}\sqrt{1-r^2}U_j(r)\sigma_m(r)dr$$

$$\hat{C}_{ij}\hat{f}_j + \hat{D}_{ij}\hat{g}_j = \frac{1}{E_s}\int_{-1}^{1}\sqrt{1-r^2}U_j(r)\sigma_b(r)dr \tag{48}$$

where U_j, $(i,\ j = 0,\ 1,\ 2, \cdots N)$, denotes the Chebyshev polynomials of the second kind.

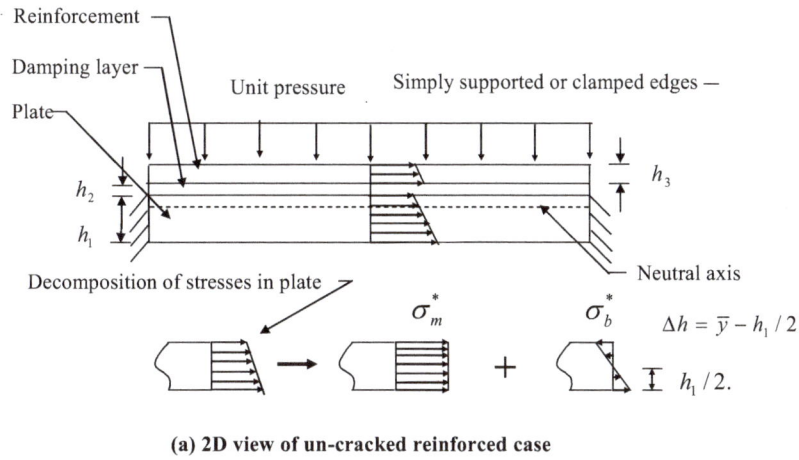

(a) 2D view of un-cracked reinforced case

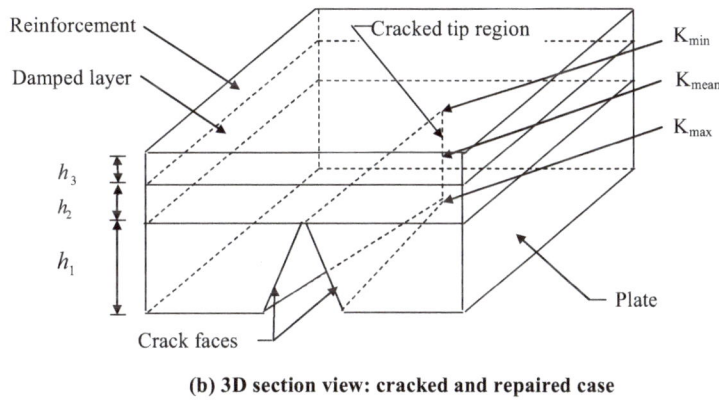

(b) 3D section view: cracked and repaired case

Figure 13: One sided repair consisting of a reinforced beam subject to one unit of lateral pressure.

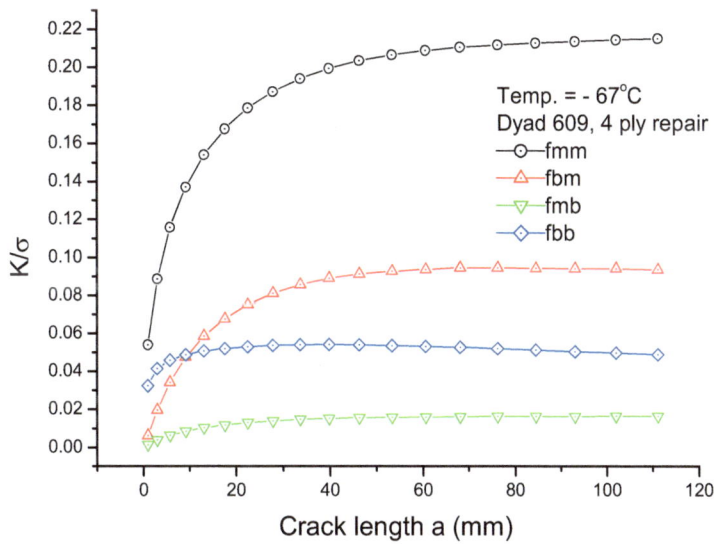

Figure 14: Results from Wang-Rose crack bridging model [23] for lateral static loading at -67° C.

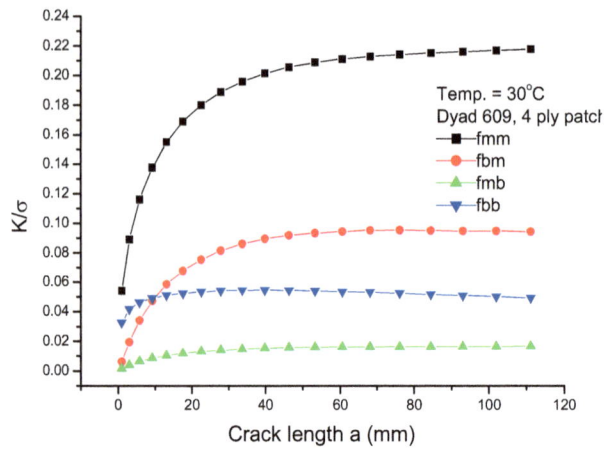

Figure 15: Results from the Wang-Rose crack bridging model [23] for lateral static loading at 30° C.

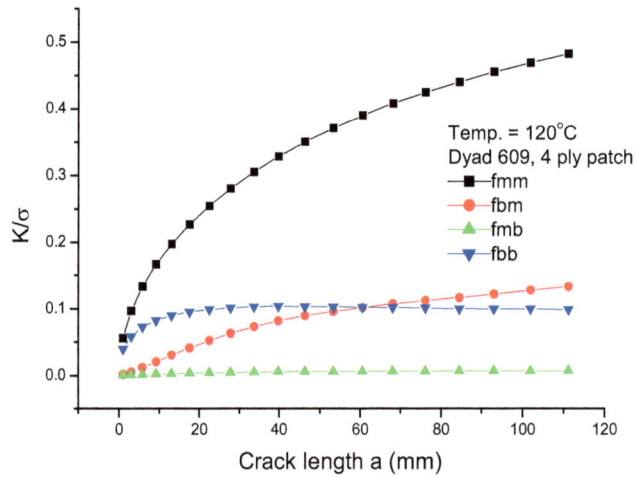

Figure 16: Results from the Wang-Rose crack bridging model [23] for lateral static loading at 120° C.

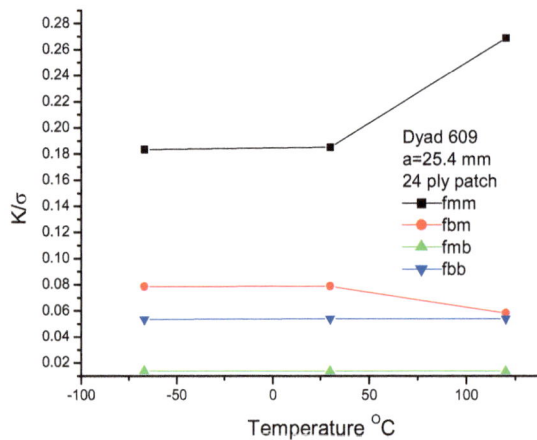

Figure 17: Results from the Wang-Rose crack bridging model [23] for lateral static loading and crack-length a= 25.4mm.

THERMAL RESIDUAL STRESS

At a given operating temperature, the thermal stresses can be considered as the superposition of the thermal stresses due to two steps: the curing of the adhesive (involving a localized heating-cooling cycle) and a uniform temperature change from the room temperature to the operating temperature. For an aircraft panel that is riveted on to stiff frames, the aircraft skin is completely restrained from expansion and contraction during the heating-cooling cycle. Consequently no thermal stress will develop as a result of the bonding of the repair patch.

During operations the aircraft skin temperature can drop to as low as -67°C. With the aircraft skin being stiffened by the repair and constrained by frames, a high thermal stress will occur in the repair and the skin. A conservative approach is to assume that all edges of the skin panel are completed constrained, i.e., the total strain remains zero. The resulting thermal stresses in the skin are given by,

$$\sigma_{xx}^T = \sigma_{yy}^T \equiv \frac{1}{1-v}\alpha_1 E_1 (RT - T_{oper}) \tag{49}$$

where RT and T_{oper} denote respectively the room temperature and the operating temperature, also the thermal coefficient of expansion of the plate is denoted by α_1. The above expression furnishes an upper-bound estimate of the thermal stresses.

A lower-bound estimate of the thermal stresses can be obtained by assuming that the frames experience the same temperature change as the skins. In this case, the problem can be viewed as a large plate containing a small patch. The thermal stress in the skin, underneath the patch, is given by the Rose and Wang [2, 27],

$$\sigma_{xx}^T = \sigma_{yy}^T \equiv \frac{(1+v_1)(1-\alpha_3/\alpha_1)S}{2(1-v_3)+(1-v_1^2)S}\alpha_1 E_1 (RT - T_{oper}) \tag{50}$$

where S denotes the stiffness ratio given by,

$$S = E_3 h_3 / E_1 h_1 \tag{51}$$

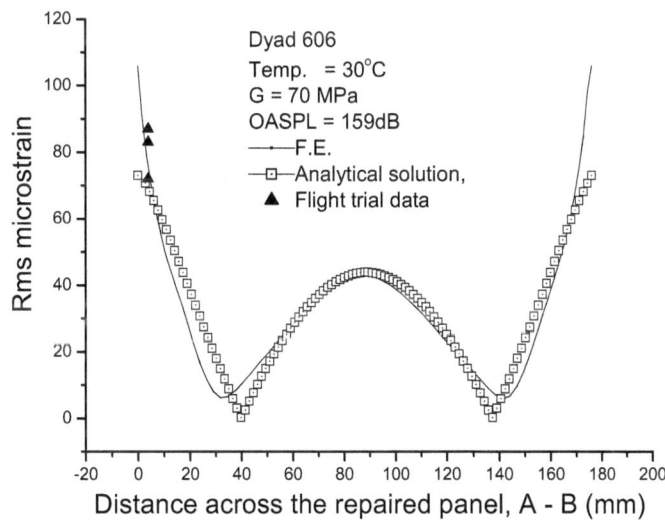

Figure 18: Distance across the repair (A – B) versus microstrain.

The above thermal stresses are membrane stresses. The stress intensity factor associated with the thermal stresses are

$$K_m^T = \sigma_{xx}^T f_{mm} \tag{52}$$

$$K_b^T = \sigma_{xx}^T f_{bm} \tag{53}$$

$$K_T = K_m^T + K_b^T \tag{54}$$

The influence of thermal stresses on fatigue crack growth rates will be discussed in a following section.

VALIDATION OF REPAIRED PANEL

The second part of the validation, now knowing the sound pressure level, was to predict the microstrain and compute the stress intensity for a repaired plate. The results are shown in Fig. **18**. It is seen that the strains measured during test flight near the edge of the plate was 87 rms microstrain for flight 1, 83 microstrain for flight 2, and 72 microstrain for flight 4. Overall these results have shown a good correlation with test flight strains.

CRACK GROWTH MODEL

A comparison of the data from standard crack growth laws [28] with acoustic data [29] showed a good agreement. The standard crack growth laws are intended to be accurate from 6 to 50 MPa$\sqrt{m}$ under in-plane tension cycling only. The threshold stress intensity factor for crack growth under high frequency out-of-plane bending may be higher [30] than for in-plane loading. The crack growth law for 7075-T6 is given by [31] as:

$$\frac{da}{dn} = C_f (\Delta K_{rms})^m \left[1 - \left(\frac{\Delta K_{th}}{\Delta K_{rms}} \right)^2 \right] \tag{55}$$

where $C_f = 2.1 \times 10^{-7}$, $m = 3.02$. Here ΔK_{th} is root-mean-square of the threshold stress intensity factor for crack growth.

CRACK GROWTH THRESHOLD

Data derived from [29] indicate that the crack growth threshold for peak values of stress intensity factor, ΔK_{th}, is approximately 10 MPa$\sqrt{m}$.

CRACK CLOSURE

It is known that under cyclic lateral loading, plasticity will occur at the crack tip [31]. As a result, during the second tension cycle there will be a delay until the crack opens. The fraction of the load at which the crack opens is defined as β. As shown in Fig. **19**, the time history for a random event in which a maximum peak occurs is defined by αK_{rms}. From [32] the effective (peak) stress intensity factor due to crack closure is given by,

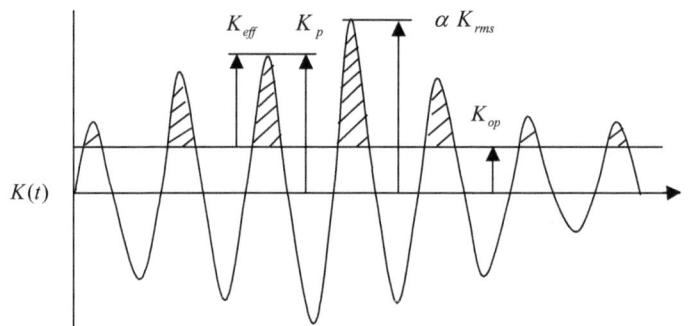

Figure 19: Effect of crack closure in random bending, variation of K with time [32].

$$\Delta K_{eff} = K_p - \alpha \beta K_{rms} \tag{56}$$

Using the values of α and β available from [30], i.e., $\alpha = 3.75$ and $\beta = 0.3$ respectively, the effective K is

$$\Delta K_{eff} = K_p - 1.125 K_{rms} \tag{57}$$

CRACK GROWTH FOR NARROW RANDOM CRACK GROWTH

Under cyclic flexural load crack faces may come into contact even when the surface of the plate is in tension, just like in the case of crack under tensile loading. This results from the plastic wake left behind an advancing crack; this plastic wake will result in crack closure before complete unloading occurs. From [31] a peak probability distribution has been used as representative of acoustic loading and exhibits a "truncated" probability distribution in which peaks above a certain level do not occur. From flight trials data [24] the ratio of peak values to rms values is 5.

Case A

If the stress intensity factor at which the crack opens ($\alpha \beta K_{rms}$) is greater than the peak value of the threshold stress intensity $\sqrt{2} \, K_{rms}^t$ then peaks less than $\alpha \beta K_{rms}$ will not contribute to the crack growth. In this case, the crack growth rate is given by

$$\frac{da}{dN} = \frac{C}{(1-\beta^2)^n} K_{rms}^n \, I(n, \frac{\alpha^2 \beta^2}{2}, \frac{\alpha^2}{2}, \frac{\alpha^2 \beta^2}{2}) \tag{58}$$

Case B

If $\alpha \beta K_{rms}$ is less than $\sqrt{2} \, K_{rms}^t$ then this will not result in crack growth, due to a threshold effect. In this case,

$$\frac{da}{dN} = \frac{C}{(1-\beta^2)^n} K_{rms}^n \, I(n, \frac{\alpha^2 \beta^2}{2}, \frac{\alpha^2}{2}, \frac{(K_{rms}^t)^2}{K_{rms}^2}) \tag{58}$$

where

$$I(n, d, h, l) = \int_l^h (t^{1/2} - d^{1/2})^n \, e^{-t} \, dt \, , \tag{59}$$

The alternative to solving these equations is to design the repair such that $K \le K_p$.

ACOUSTIC LOADING ON F/A-18

While data exist [3] for the SPL on the inlet nacelle region, there is no data for the aft fuselage region. It was reported that the airframe [3] was designed for an overall sound pressure level (OASPL) of 160dB. Flight trials [24] in which a dosimeter was installed to measure the strains on two adjacent skin panels with and without repair, respectively, have provided both peak and rms strain data on the aft fuselage region. For the un-repaired panel two strain gauges were used, with one gauge at the centre of the panel and the other at the edge of the panel. Since the centre gauge was in a region of low strain gradient, the measured strains are more reliable than the gauge close to the edge. Assuming that the mode shape is the fundamental mode, it is possible to back-calculate the OASPL from the centre strain gauge data. These calculations show that the OASPL in the region of the aft fuselage is 159 dB, very close to the design level of 160dB.

HIGHLY DAMPED REPAIR EXAMPLE

An example problem such as a highly damped repair for the aft fuselage of the F/A-18 will be considered. Data needed for this design are shown in Tables **3**, **4** and **5**. The various parameters used in the analysis are shown in Table **3**. The expected operating temperature of the F/A-18 in the vicinity of the aft fuselage is -67 to +120°C.

Consider a plate of dimensions $L_x = 176\,mm$, $L_y = 175\,mm$ and thickness $h_1 = 0.9$ mm. The repair patch has a thickness $h_3 = 0.508$ mm, which corresponds to four plies of boron composite as the constraining layer shown in Fig. **4**. The aim is to place the overall neutral axis within the viscoelastic material in order to maximize the shear. Considering an operating temperature of –40 to 90°C, Dyad 609 is used, in this case due to durability concerns of Dyad 606 at high temperature. Dyad 609 material only comes in at thicknesses of 0.508 and 1.106 mm. A thickness of 0.508 was chosen to minimise changes to the external mouldline. For the boron doubler, a stacking sequence of [0/90/0/90] will be used, resulting in a laminate modulus of 114 GPa. Although the laminate is unbalanced, the bending stiffness is the same in both x and y directions.

Firstly, from experimental data [31], both the loss factor and the resonant frequency change with increasing moisture content. Note that the un-conditioned specimen gives higher loss factors than data provided by the manufacturer. The difference is due to moisture uptake whereas the manufacturer's data is for low moisture content. A change of resonant frequency indicated that the shear modulus also changed with moisture uptake. These are factors that may have to be considered in determining the real loss factor. For this exercise the manufacturers data will be used, see Fig. **20**.

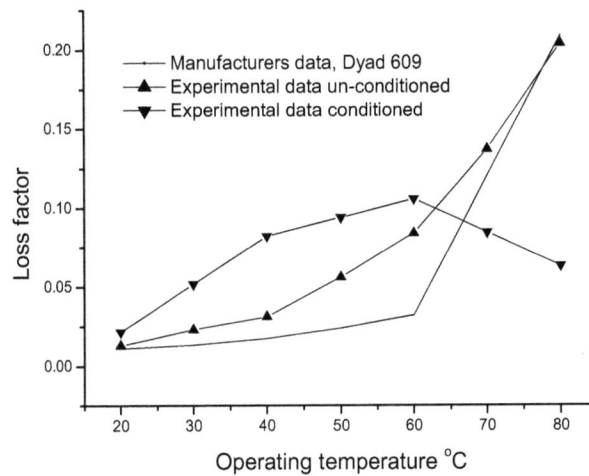

Figure 20: Loss factor versus operating temperature for Dyad 609 manufacturers data, un-conditioned and conditioned specimens with the 4 ply constrained layer configuration.

With the parameters given in Table **3**, the design conditions can be solved numerically or symbolically using Mathematica™. The stress intensity factors can then be computed for a range of operating temperatures and SPL's. The corresponding PSD of the excitation is tabulated in Table **6** for a plate with and without repair. The material properties of Dyad 606 and 609 are listed in Table **4** and **5** respectively, for various operating temperatures. These results are graphically displayed in Figs. **21** and **22** for OASPL of 159dB and operating temperatures from –67 to 120°C.

Table 3: Details of parameters used in design.

Item	Value
h_1	0.9 mm
h_2	0.508 mm
h_3	0.508 mm
E_1	71000 MPa
E_3	113300 MPa
v_1	0.3
v_3	0.0535
ρ_1	2.77×10^{-9} kkg/mm^3
ρ_3	2.03×10^{-9} kkg/mm^3

ρ_2	1.10×10^{-9} kkg/mm^3
L_x	175 mm
L_y	176 mm
m	3.02
C_f	1.37×10^{-5}
K_f	63.9
η_2	1.0 *
η_1	0.013
η_3	0.004
G_2	10 MPa*

Table 4: Loss factor and shear modulus for Dyad 606 versus operating temperature.

Operating temp. ^{O}C	Loss factor $\eta + \eta_t$	G_2 (MPa)
-67	0.01912	300
-40	0.01912	300
-20	0.01912	300
0	0.01912	300
20	0.02025	300
30	0.1749	70
40	0.4171	17
50	0.3619	5
60	0.1385	2
70	0.04151	.7
80.	0.02391	.4
90	0.02146	.4

Table 5: Loss factor and shear modulus for Dyad 609 versus operating temperature.

Operating temp. ^{O}C	Loss factor $\eta + \eta_t$	G_2 (MPa)
-67	0.02109	300
-60	0.02109	300
-40	0.02109	300
-20	0.02109	300
0	0.02109	300
20	0.03224	300
30	0.04088	280
40	0.05717	200
50	0.09915	130
60	0.14563	85
70	0.23625	50
80.	0.36454	25
90	0.40449	5
100	0.19466	1.7
110	0.13307	1.0
120	0.06869	0.9

Table 6: PSD of the excitation of the plate alone and also the repaired case.

OASPL (dB)	S_l (197 Hz)	S_l (344 Hz)
159	2.196×10^{-9}	2.015×10^{-9}

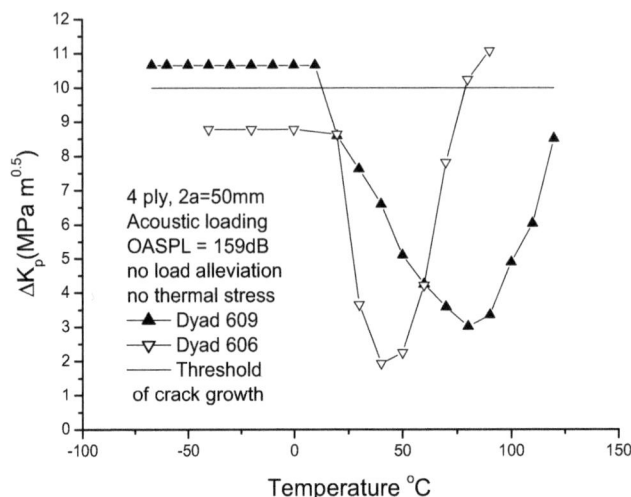

Figure 21: Predicted peak stress intensity factor versus operating temperature at OASPL=159dB, using Dyad 609 or 606, for crack length 2a=50mm, with a 4 ply constrained layer configuration where thermal stresses due to uniform temperature change have been ignored.

In Fig. **22** the peak microstrain for Dyad 606 and 609 is 2000 and 955 respectively. Note that the strain level of 1750 microstrain corresponds to the un-patched panel derived from flight trials results. Cleary the Dyad 609 patch has reduced the strain by a factor of almost 2. Note that the Dyad 606 design increases the maximum strain at high temperatures to a level exceeding the flight trial strains on the un-patched panel. This is the result of the damping material having a very low shear modulus at higher temperatures, with the additional mass of the damped layer increasing resonant response. Also shown in Fig. **21** is the threshold value for crack growth. As discussed, the threshold value of 10 MPa m$^{0.5}$ was determined from an experimental study [30]. It is evident that from Fig. **21** that the 4 ply constrained layer Dyad 606 design will not survive the high temperature loading. Also the 4 ply constrained layer Dyad 609 is just above the threshold at low temperatures. However it is expected that when using Dyad 609, a six-ply constrained layer configuration will suffice.

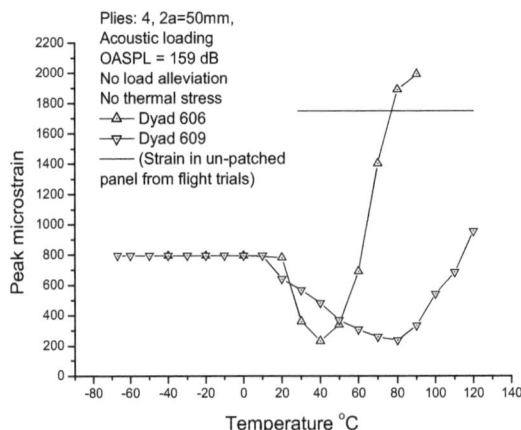

Figure 22: Predicted peak microstrain for repaired panel versus operating temperature for OASPL=159dB, using either Dyad 609 or Dyad 606, with a 4 ply constrained layer configuration.

ADHESIVES FOR THE REPAIR

Further considerations concerning the real operating temperatures were made in [32]. For the coldest day the entire aircraft skin temperature is -67°C. For the hottest day of the year the repair was adjacent to both 103°C and 141°C temperature regions. As a result an average hottest day was taken to be 122°C, and the analysis was carried out at 120°C. The adhesive used to construct the patch was FM300, which was co-cured with the boron and damping material at 177°C. The adhesive used to bond the repair to the aircraft skin was FM300-2 since this adhesive would perform within the range of -67°C to 120°C. Figs. **23** and **24** show schematically the structural components of the patch system, and the thickness of each component. In this section adhesive shear stresses due to an acoustic loading will be evaluated.

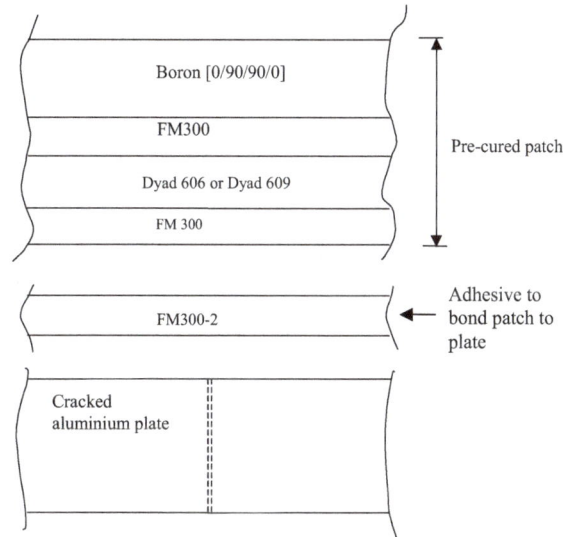

Figure 23: Through thickness view of the individual structural components for the repair of the plate.

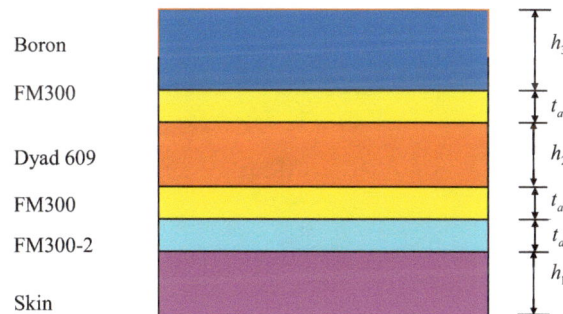

Figure 24: Diagram showing thicknesses of patch components.

The shear stress in the adhesive due to acoustic loading, which induces membrane and bending stresses (σ_m and σ_b), is given by,

$$\tau = \frac{G_a^* l}{t_a^*}\left[\frac{\sigma_m + \sigma_b}{E_1}\right],$$

(60)

where

$$\frac{t_a^*}{G_a^*} = \left(\frac{t_{a1}}{G_{a1}} + 2\frac{t_{a2}}{G_{a2}} + \frac{h_2}{G_2}\right),$$

(61)

$$l = \left[\left(\frac{t_{a1}}{G_{a1}} + \frac{2t_{a2}}{G_{a2}} + \frac{h_2}{G_2} \right) \frac{E_1 E_3 h_1 h_3}{(E_1 h_1 + E_3 h_3)} \right]^{1/2} \tag{62}$$

The parameters G_{a1}, G_{a2} and G_2 denote the adhesive shear moduli for FM300-2, FM300, and the damping layer respectively. The Young's moduli of the skin and the patch are denoted respectively as E_1 and E_3. Parameters h_1, h_2 and h_3 denote the thickness of the skin, the damping layer, and the patch respectively. The thickness of the FM300-2 and FM300 adhesives are t_{a1} and t_{a2} respectively.

The shear modulus of the adhesive varies with temperature, see Fig. **25** for the stress strain curve.

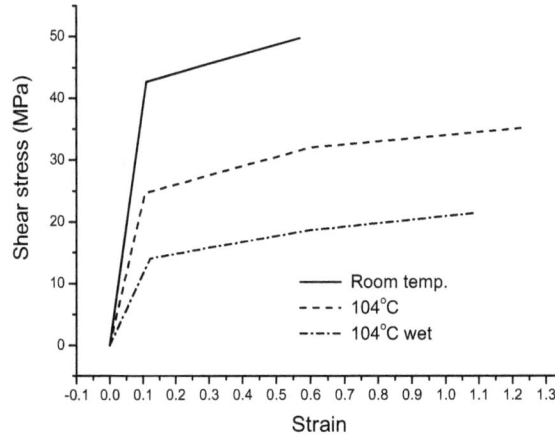

Figure 25: Stress strain diagram for FM300 for the temperature range shown, [34].

Calculations in Tables **7** and **8** for acoustic shear stresses in the adhesive have confirmed that at an operating temperature of 0 or 120^0C, both FM300 and FM300-2 are operating within the elastic region, hence no fatigue damage will occur, see Table **8**.

Table 7: Parameters for adhesive shear stress.

Temp. ^{0}C	G_{a1} (MPa)	G_{a2} (MPa)	G_2 (MPa)
120	199	199	1
0	415	415	300

$(t_{a1} = t_{a2} = 0.127 \, mm, h_3 = 0.762 \, mm)$

Table 8: Adhesive shear stresses due to acoustic loading using parameters from Table **7**.

Temp ^{0}C	σ_m (MPa)	σ_b (MPa)	$^* \tau_P$ (MPa)	$^{**} \tau_{peak}$ (MPa)
120	0.544	14.662	14.	0.304
0	6.621	6.036	43.	3.540

*is the plastic shear stress for the adhesive.

**is the peak value $\tau_{peak} = 5\tau$

ALTERNATING STRESS INTENSITY FACTOR

The effect of low temperatures may result in tensile residual stresses which would keep the crack open, thus increasing crack growth rates. In this case a simple and conservative approach is to assume that negative stress intensity factor does not contribute to crack growth. In this case, the effective crack growth driving force can be expressed as

$$\Delta K_{eff} = \min(K_{max}, K_{max} - K_{min}), \ K_{max} > 0 \tag{65}$$

Under acoustic loading, the minimum stress intensity factor K_{min} is equal in magnitude to the maximum stress intensity factor but opposite in sign, *i.e.* $K_{min} = -K_{max}$. In the presence of thermal residual stress intensity factor K_T, the effective range of stress intensity factor is:

$$\Delta K_{eff} = 2K_{max}, \qquad K_T > K_{max} \tag{66a}$$

$$\Delta K_{eff} = K_{max} + K_T, \quad K_T \leq K_{max} \tag{66b}$$

A conservative assumption for the $K_T \leq K_{max}$ case is:

$$\Delta K_{eff} = K_{max} \tag{67}$$

SMALL PATCH

When a small patch is applied, due to the lack of constraint, high residual thermal stress will exist in the skin after the bonding of a Boron patch. It has been found that the thermal stress intensity factor is approximately equal to the peak stress intensity factor at 20°C. Therefore, the effective range of the stress intensity is:

$$\Delta K_{eff} = K_{max} + K_T \approx 2K_{max} \tag{68}$$

This means that small patches will have large thermal residual stresses and may not prevent crack growth. This is consistent with the previous failed-attempt of repairing the acoustic fatigue crack with a small Boron patch that covered only the cracked region.

LARGE PATCH

For a large patch that covers the entire bay (panel) that is rigidly connected to the airframe, no thermal expansion or contraction would occur during the bonding process. As a result, the repaired skin will not experience any thermal residual stress, *i.e.,* $K_T = 0$.

$$\Delta K_{eff} = K_{max} \tag{69}$$

It is clear that the effective crack growth driving force would be much less than that pertinent to a small patch.

At low temperatures thermal mismatch may result in a high thermal stress intensity factor for the combination of aluminium and boron. In this case, the thermal stress intensity factor may be greater than that induced by the acoustic loading, i.e. $K_T \geq K_{max}$. As result during each cycle, the effective stress intensity factor range $\Delta K_{eff} = 2 K_{max}$.

OPERATING TEMPERATURES

As mentioned previously the maximum and minimum temperatures in this area are –67°C and 120°C respectively. The –67°C temperature corresponds to the lowest airspeed and engine power that will maintain the aircraft in straight and level flight at Flight Level (FL) 500. Under this condition the maneouver capability is limited. The maximum temperature corresponds to Mach 2 at FL 350. While the maneuver capability for the coldest temperature is limited, it is not necessary to use a load alleviation factor. Note that the relationship of shear modulus to temperature for FM300 is:

$$G = 415 - 1.8T \qquad (70)$$

The damping material, Dyad 609, is most effective over the range of temperature 20-120°C. At low temperatures the loss factor of the repair will only be 0.004. Available data indicates that the properties of Dyad 609 do not change at low temperatures. Hence at low temperatures the repair is not damped. The patch works as a standard patch repair at low temperatures, except that the damping layer increases the bending stiffness above that expected from a repair without the damping layer. This in turn reduces the stress intensity and hence crack growth.

RESULTS

Firstly consider an un-damped patch in Fig.26. This is a classic example that occurs when un-damped patches are used. One repair is a simple un-damped boron patch repair while the other is an un-damped aluminium patch. Both repairs result in stress intensity factors exceeding fatigue growth threshold. The patches are bonded to the skin, but the adhesive softens with rise of operating temperature, thus resulting in low repair efficiency.

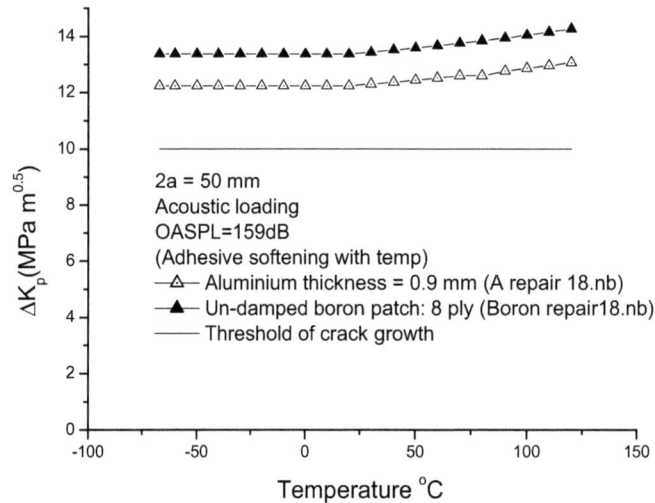

The plot legend reads:

2a = 50 mm
Acoustic loading
OASPL=159dB
(Adhesive softening with temp)
—△— Aluminium thickness = 0.9 mm (A repair 18.nb)
—▲— Un-damped boron patch: 8 ply (Boron repair18.nb)
—— Threshold of crack growth

The vertical axis is $\Delta K_p \ (\mathrm{MPa\ m^{0.5}})$ and the horizontal axis is Temperature °C.

Figure 26: Predicted peak stress intensity factor versus operating temperature for un-damped boron repair, 8 plies, and un-damped aluminium plate repair.

The performances of damped repair incorporating patches made of four, six, and eight plies are presented in Fig. **27a** for acoustic loading. Also included in the figure are the predictions of the closed-form solutions for four-ply patch, indicating a good agreement. These results suggest that both the 6 and 8 ply designs would be able to completely arrest fatigue cracks less than 50mm long.

It is known that as the crack grows the stress intensity in a panel, surrounded by structure, will decrease as the crack tip approaches the boundary [13]. The present analysis is capable of predicting the change in stress intensity as the crack grows as the prospective membrane and bending stresses in equation (48) decrease as the crack tip approaches panel edge.

The results for a crack size of 2a=75mm are shown in Fig. **27b**. All three designs with four, six, and eight ply-patch are effective under this loading. As the crack length extends to 2a=100mm, as shown in Fig. **27c**, and 2a=150mm in Fig. **27d**, all three designs remain effective. It is evident that the six-ply patch design is always below the threshold for crack growth. Hence a six-ply constraining layer design is capable of stopping very large cracks. Overall the results show that the analytical solution underestimates the F.E. results slightly at low temperatures and a safety factor 1.05 is needed. At high temperatures, a factor of 1.2 is required.

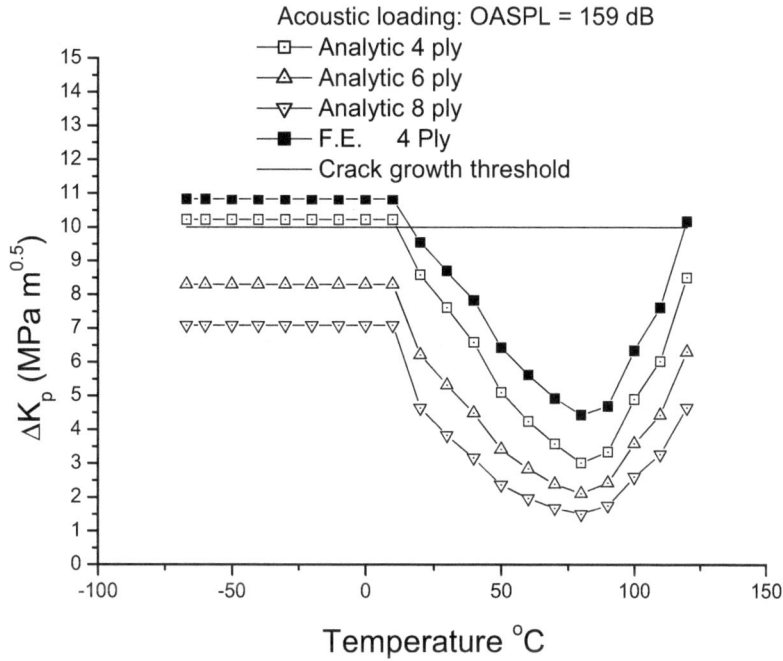

Figure 27(a): Predicted peak stress intensity factor versus operating temperature at OASPL=159dB, using Dyad 609, for crack length 2a=50mm, with 4, 6 and 8 ply constrained layer configurations.

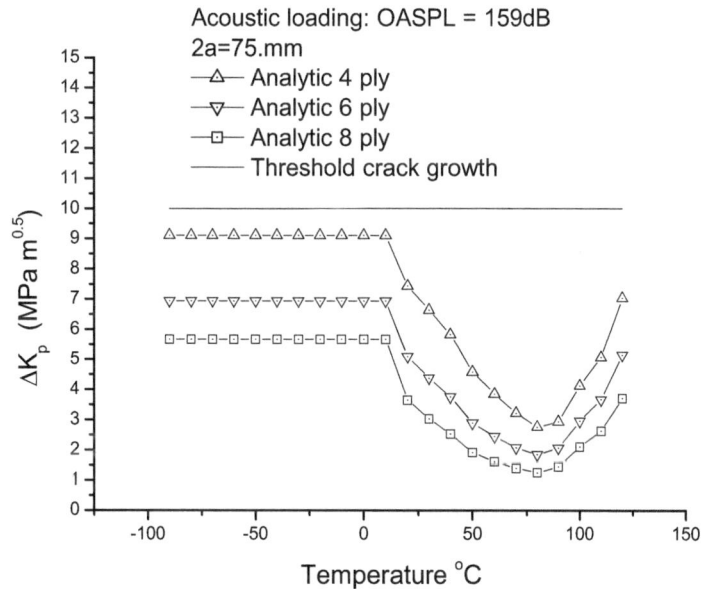

Figure 27(b): Predicted peak stress intensity factor versus operating temperature at OASPL=159dB, using Dyad 609, for crack length 2a=75mm, with 4, 6 and 8 ply constrained layer configurations.

Shown in Fig.28 is the variation in predicted rms microstrain versus temperature for three different configurations. The six-ply patch design is the optimum design over the required operating temperature.

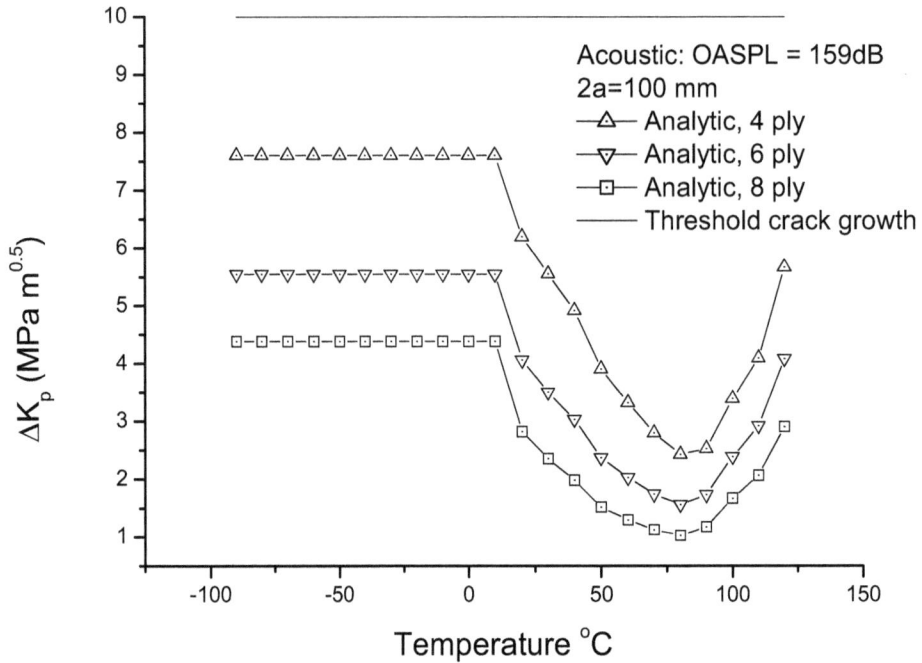

Figure 27(c): Predicted peak stress intensity factor versus operating temperature at OASPL=159dB, using Dyad 609, for crack length 2a=100mm, with 4, 6 and 8 ply constrained layer configurations.

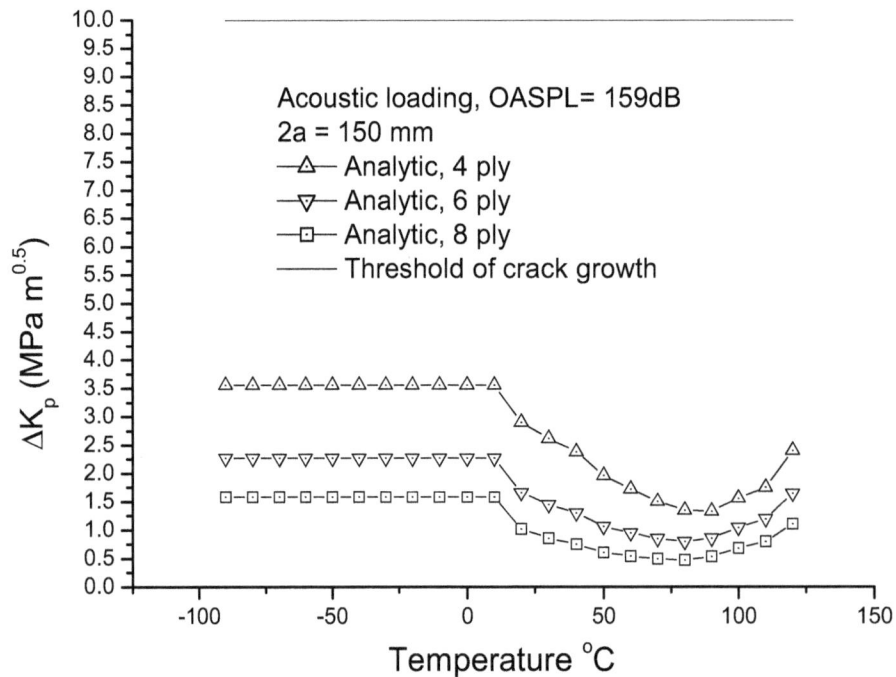

Figure 27(d): Predicted peak stress intensity factor versus operating temperature for damped repairs using Dyad 609 for 4, 6 and 8 ply constrained layer lay-ups. The crack length is 2a=150mm.

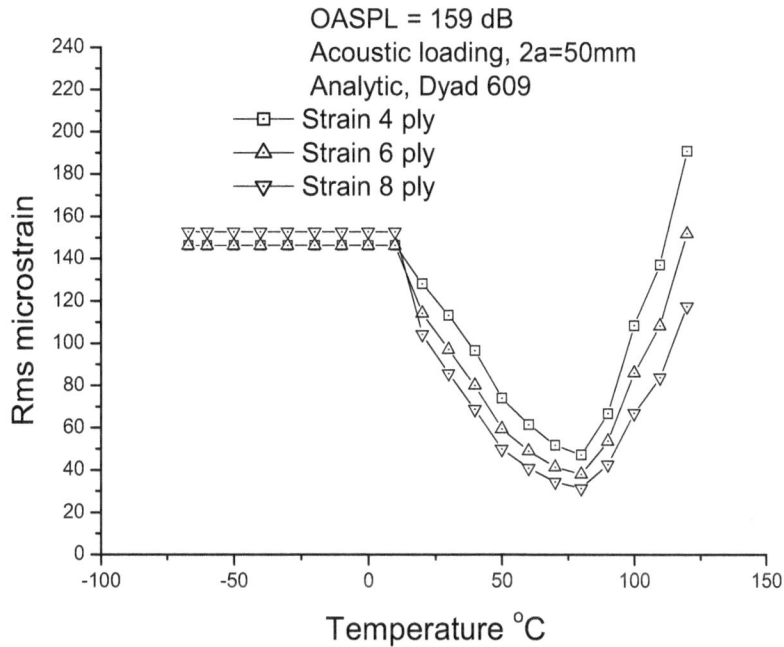

Figure 28: Residual predicted rms strain versus operating temperature at OASPL=159dB, using Dyad 609, for crack length 2a=50mm, with 4, 6 and 8 ply constrained layer configurations.

CONCLUSIONS

A generic design procedure has been developed using a number of closed form solutions. These consist of a simplified dynamic analysis together with crack bridging theory. Furthermore, the solution is also applicable for plates containing cracks near boundaries, accounting for the effect of residual thermal stresses. The procedure uses crack growth equations which have been found to give good results for acoustic fatigue conditions. It has been shown that the analysis for crack bridging can predict the repair efficiency with all the plate edges being either simply supported or clamped. As a result, this generic design procedure will enable optimum design of highly damped repairs to acoustically-induced cracked panels. It has been found that a highly damped repair is a viable, low risk, low cost repair to prevent any crack growth on the aft fuselage of a typical fighter aircraft.

NOMENCLATURE

CLD	=	Constrained layer damping
C_f	=	Coefficient for crack growth law
D	=	Bending stiffness of plate
D_t	=	Overall bending stiffness of 3 layer plate
E	=	Young's modulus
E_1	=	Young's modulus for plate
E_1'	=	Real part of complex Young's modulus for plate

E_3 =   Young's modulus for constraining layer

E_3' =   Real part of complex Young's modulus for constraining layer

F =   Geometry factor

G_2 =   Shear modulus for damping layer

G_a =   Shear modulus for adhesive

G_2' =   Real part of complex shear modulus

I_1 =   Second moment of area of layer 1

I_3 =   Second moment of area of layer 3

I_t =   Total bending stiffness

I_{tr} =   Real part of total bending stiffness

L_x =   Length of plate in x direction

L_y =   Length of plate in y direction

K =   Stress intensity factor

K_p =   Peak stress intensity factor

K_{min} =   Minimum stress intensity factor

K_{mean} =   Mean stress intensity factor

K_{max} =   Maximum stress intensity factor

K_{rms} =   Root mean square stress intensity factor

K_m =   Membrane stress intensity factor

K_b =   Bending stress intensity factor

K_m^{rms} =   Root mean square membrane stress intensity factor

K_b^{rms} =   Root mean square bending stress intensity factor

$\overline{K}$ =   Stress intensity factor due to application of unit pressure on structure

$K_{rms\text{-}u}$ =   Root mean square stress intensity factor un-repaired

$K_{rms\text{-}r}$ =   Root mean square stress intensity factor repaired

K_{th} =   Threshold stress intensity factor for crack growth

K_{eff} =   Effective stress intensity factor

$(K_{max})_{rms}$	=	Maximum rms stress intensity factor
K_T	=	Stress intensity factor due to thermal stresses
K_m^T	=	Stress intensity factor due to thermal membrane stresses
K_b^T	=	Stress intensity factor due to thermal bending stresses
OASPL	=	Overall Sound Pressure Level
PSD	=	Power Spectral Density
RT	=	Room temperature
SPL	=	Sound pressure level
S_o	=	Power spectral density of the response
S_I	=	Power spectral density of the excitation
$S_I(f_0)$	=	Power spectral density of the excitation at frequency f_0
ΔT	=	Change from cure to operating temperature
T_c	=	Cure temperature
T_{oper}	=	Operating temperature
Y	=	Geometric factor
W	=	Geometry factor for frequency
a	=	Half crack length
d_1	=	Distance between centroid of section and centre of plate
d_3	=	Distance between centroid of section and centre of constraining layer
f	=	Vibration frequency of plate
f_m	=	Membrane component of the
	=	normalized stress intensity factor
f_b	=	Bending component of the
	=	normalized stress intensity factor
f_o	=	First resonant frequency of the un-repaired plate
f_c	=	First resonant frequency of the repaired plate
Δf	=	Frequency range

g_{opt}	=	Optimum value of shear parameter
g_m	=	Geometric shear parameter
g'	=	Real part complex geometric shear parameter
g^*	=	Complex shear parameter
h	=	Thickness of plate
h_1	=	Thickness of plate for layer 1
h_2	=	Thickness of plate for layer 2
h_3	=	Thickness of plate for layer 3
h_{13}	=	Distance between centre of layer 1 and centre of layer 3
k_f	=	The wave number
k_x	=	Modal wave number in x direction
k_y	=	Modal wave number in y direction
m_t	=	Total mass per unit length
m	=	Fundamental mode in x direction
n	=	Fundamental mode in y direction
p_{rms}	=	rms fluctuating pressure
s	=	Stiffness ratio
t_a	=	Thickness of adhesive layer
w	=	Transverse displacement of plate
α_1	=	Thermal coefficient of expansion for layer 1
α_3	=	Thermal coefficient of expansion for layer 3
ε_b	=	Bending strain
ε_m	=	Membrane strain
β	=	Fraction of peak load that crack opens
η	=	Loss factor
η_s	=	Loss factor for damped plate
η_c	=	Loss factor for viscoelastic damping layer

$\eta_{s,\max}$	=	Optimum loss factor for 3 layer plate
η_1	=	Loss factors for layer 1
η_2	=	Loss factors for layer 2
η_3	=	Loss factors for layer 3
η_t	=	Non-core loss factor
$\eta_{DC,m}$	=	Modal loss factor
ν	=	Poissons ratio for plate
ν_1	=	Poissons ratio for plate component 1
ν_3	=	Poissons ratio for plate component 3
σ_b	=	Far field bending stress
σ_m	=	Far field membrane stress
σ_{xx}^T	=	Thermal stress in x direction
σ_{yy}^T	=	Thermal stress in y direction
τ_p	=	Plastic shear stress for adhesive
τ_{peak}	=	Peak shear stress for adhesive
ρ	=	Density of plate
ρ_1	=	Density of plate layer 1
ρ_2	=	Density of damping layer, layer 2
ρ_3	=	Density of constraining layer, layer 3
ω_M	=	m[th] resonant frequency
w	=	Transverse displacement
w_o	=	Transverse displacement at centre of panel
Ω	=	Non-dimensional frequency ratio

REFERENCES

[1]   Baker AA, Jones R (Eds), Bonded Repair of Aircraft Structures. Martinus Nijhoff Publishers, The Hague, 1988.

[2]   Duong C N. and Wang CH, Composite repair: theory and design, Elsevier, 2007.

[3]   Brewer TK., A/B/C/D Aircraft Lower Nacelle Skin Acoustic and Strain Measurements and Sonic Fatigue Analysis. McDonnell Douglas Aerospace, MDC 94B0044, Mar, 1994.

[4]   Callinan RJ, Galea SC, Sanderson S, Finite element analysis of bonded repairs to edge cracks in panels subjected to acoustic excitation. J. Comp. Struct. 1997; 649-660.

[5]   RAAF Engineering Standard DEFAUST9005-A.

[6]   Callinan RJ, Chiu WK and Galea SC, Optimization of a Composite Repair to Cracked Panels Subjected to Acoustic Excitation. Paper A98-31631 21st ICAS Congress 13-18 Sept. 1998; Melbourne, Australia.

[7]   Callinan RJ, Wang CH, Galea SC, Sanderson S, Rose LRF, Analytical Solution for the Stress Intensity Factor in Bonded Repairs to Panels Subjected to Acoustic Excitation. Presented to joint FAA/DoD/NASA Conference on Aging aircraft. 20-23 Sept. 1999.

[8]   Callinan RJ, Wang CH, Sanderson S, Galea SC. Generic Design Procedures for the Repair of Acoustically Damaged Panels. DSTO-TR-0283, AR-013-231, June 2008.

[9]   Rogers L, Maly J, Searle IR, Begami RI, Owen W, Smith D, Gordan RW, Conley D, Durability Patch: Repair and Life extension of high-cycle fatigue damage on secondary structure of ageing aircraft, 1ST Joint DOD/FAA/NASA Conference on Aging Aircraft, Ogden, Ut, 8-10 July, p595-623, 1997.

[10]   Liguore SL, Hunter K, Perez R, Beier TH, Flight Test Evaluation of Damped Composite Repairs For Sonic Fatigue. Sonic Fatigue Session of the 40th AIAA/ASME/ASCE/AHS SDM Conference St. Louis, MO April 12-15, p1498-1508, 1999.

[11]   Gordon RW, Hollkamp JJ, Liguore S, Callinan RJ, An Experimental Investigation of Damped Repairs for Sonic Fatigue. Fifth Joint NASA/FAA/DoD Conference on Aging Aircraft, Sept. 2001.

[12]   Liguore SL, Beier TH, Gordon RW, Hollkamp JJ, Design and Analysis of Damped Repairs For Sonic Fatigue Cracking. The 2001 USAF Aircraft Structural Integrity Program Conference, 11-13 Dec. Williamsburg, Virginia, USA.

[13]   Ikegami R, Haugse E, Trego A, Structural Technology and Analysis Program. AFRL-VA-WP-TR-2001-3037, June 2001.

[14]   Mead DJ, Markus S. The forced vibration of a damped sandwich beam with arbitrary boundary conditions. J.Sound Vib., 1969; 10: 2.

[15]   Mead DJ, Passive Vibration Control. John Wiley and Sons, 1982.

[16]   Byrne KP, Bending Induced Crack Propagation in Bare and Clad 4% Cu-Aluminium Alloys with Reference to Acoustically Propagated Fatigue Cracks. Institute of Sound and Vibration Research. University of Southhampton, Technical Report No. 61, Aug. 1973.

[17]   Anon. Soundcoat data sheets for Dyad 606 and 609.

[18]   Climent H, Casalengua J, Application of a PSD technique to acoustic fatigue Stress Calculations in Complex Structures. Symposium on 'Impact of Acoustic Loads on Aircraft Structures', Paper 12, AGARD-CP-549, Lillehammer, Norway, May 1994.

[19]   Clarkson BL, Stresses in Skin Panels Subjected to Random Acoustic Loading. The Aeron. Jour., 1968; 72(695): 1000-1010.

[20]   Miles JW, On Structural Fatigue Under Random Loading. J. Aeron. Sci., 1954; 21(11): 753-762.

[21]   Byrne KP, On the Growth Rate of Bending Induced Edge Cracks in Acoustically Excited Panels. J.Sound and Vibration, 1977; 53(4) 505-528.

[22]   Young WC, Roark's Formulas for stess and strain. Sixth Edition, McGraw-Hill, 1989.

[23]   Wang CH, L.R.F Rose LRF, A crack bridging model for bonded plates subjected to tension and bending. Int. J. of Solids and Structures, 1999; 36: 1985-2014.

[24]   Callinan RJ, Ferrarotto P, Geddes R, Stoyanokski I,.Tejeddor S, Flight Trial of Dosimeter on F/A-18 to Determine the Acoustic Environment and the Effectiveness of the Highly Damped Patch. DSTO-RR-0284, AR 013-232, Oct. 2006.

[25]   Rose LRF, Theoretical Analysis of Crack Patching, in Bonded Repair of Aircraft Structures, A.A. Baker and R. Jones (eds.), Martinus Nijhoff Publishers, The Hague, 1988.

[26]   Wang CH, L.R.F.Rose, Callinan RJ, Analysis of Out-of-Plane Bending in One-Sided Repair. Int. J. Solids Structures, 1998; 35(14): 1653-1675.

[27]   Rose LRF, Wang CH, Theoretical analysis of composite repairs, Chapter 7, Advances in the bonded composite repair of metallic airframe structures, Elsevier Science Publications, London, p.137-175

[28]   Schwarmann L, Material Data of High-Strength Aluminium Alloys for Durability Evaluation of Structures. Aluminium-Verlag, Dusseldorf, 1988.

[29]   Byrne KP, Strains Affecting the Growth Rate of Edge Cracks in Acoustically Excited Panels. ISVR Tech. Rep. 59, Nov, 1972.

[30]   Callinan RJ, Wang CH, Rider A, Tejedor S, Experimental Work carried out to Evaluate the Effectiveness of Highly Damped Repairs. DSTO-TN-0589, AR-013-234, 2005.

[31]   Byrne KP. Bending Induced Crack Propagation in Bare and Clad 4% Cu-Aluminium Alloys with Reference to Acoustically Propagated Fatigue Cracks. Institute of Sound and Vibration Research University of Southhampton, Technical Report No. 61, Aug. 1973.

[32]   Barlow S, Determination of F/A-18 Thermal Profile, AR-009-066, July 1995.

[33]   Anon, FM300 Film adhesive. Manufactures data. Cytec Fiberite

CHAPTER 3

Aerothermal and Structural Dynamic Analysis of High-Speed Flight Vehicles

S.Y. Ho

Defence Science and Technology Organisation, P.O. Box 1500, Edinburgh, SA 5111, Australia

Abstract: This chapter gives a brief overview of the state-of-the-art and insight into the key aerothermal issues and challenges for aero-thermal-structural analysis and failure assessment of modern high-speed aerospace vehicles. Approximate engineering methods and computational methods to numerically simulate and understand the physics of high-speed flows surrounding the vehicle and hence, predict aerodynamic heating, the pressure and shear forces, and also the structural response to the aeroheating, aerodynamic and aeroacoustic loads are discussed in the context of the concerns of the aerospace vehicle designer and service life / structural failure analyst. The discussion is focused on those predictive methodologies that can be readily implemented, adequately robust and suitable for somewhat more complex vehicle designs and missions. Example applications, including a real-life application, are presented. Future directions, in particular the analytical / modelling challenges in the use of smart structures (e.g., SMAs for structural response / shape control), active cooling concepts and exotic composite materials for high-speed / performance aerospace vehicles are also discussed.

1. INTRODUCTION

Modern aerospace vehicles are required to operate at environments which are becoming more and more severe, for example high Mach numbers in the atmosphere for extended flight time, high-g maneuvers at the edge of the flight envelope, and so on. As a result, these vehicles (such as hypersonic systems, high supersonic aircraft and advanced missile systems) will experience very demanding heating and aerodynamic, as well as, aeroacoustic loads. It is well known that the aerothermal environment of high-speed aerospace vehicles is complex, the flow field around the vehicle can vary from continuum, transitional and rarefied depending on the velocity and altitude [1]. During the acceleration or deceleration of hypersonic vehicles the shock waves sweeping across the vehicle can interact with local shocks (shock-shock interaction) and or boundary layers (shock-boundary layer interaction) resulting in intense highly localized heating and pressure loads in the leading edges [1-8].

Severe aerodynamic heating in high-speed flights is currently one of the greatest challenges facing the R&D of future hypersonic vehicles. The high-temperature environment could cause melting and /or deformation (for example, the thermal stresses may induce buckling) of the structural components. When designing high-speed flight vehicles, it is also important to understand under what aerodynamic conditions the structure will be stable and unstable. This is more imperative at high-speeds where the structure is inherently more flexible (lower material stiffness) and unstable operation can lead to oscillations of the structure. Hence, another implication of aerodynamic heating is aerothermoelasticity effects, which usually lead to fatigue failure. In the worst case, catastrophic failure may occur if under some condition the structure is absorbing energy continuously.

An accurate prediction of the aerothermal loads, temperature distribution, structural deformations and stresses is critical to the design of high-speed vehicles, for example, in the selection of materials for construction, wall thickness, and so on. The predictive tools must be capable of predicting the multi-disciplinary aspects of aerodynamic heating and its consequences, ranging from yielding and melting of structural components, structural deformation, such as buckling and distortion of engine throat [7-8, 40] to changing the aerothermoelastic behaviour of the vehicle [33]. Some of the challenges that must be considered in the aerothermo-structural analysis include 1) the need to understand the complex operational environment, which depends on the flight trajectory of the vehicle, 2) non-linear structural behaviour arising from geometrical and /or material non-linearities (accentuated by thermal induced stresses from aerodynamic heating), 3) non-linear aerodynamics due to non-uniform flow behaviour at high-supersonic and hypersonic speeds. Furthermore, modern air vehicles are made of light weight composite materials which are inherently more flexible and aerothermoelasticity effects may be expected to become more significant at high velocities.

Predictive methods for aerothermal, structural and vibrational analysis of high speed vehicles have an important role in the design of high-speed vehicles, as well as in service life assessment and failure analysis of vehicles subjected to aerodynamic heating. The role of accurate/reliable simulation of the heating effects on structural performance is particularly important in the high supersonic / hypersonic flight regime because the testing of small / sub-scale wind tunnel models, a conventional practice in subsonic and supersonic flow, is not feasible because of the short testing time of the impulse type ground testing facilities. Aerothermal-structural and aerothermoelasticity analysis is an interdisplinary problem, where the simultaneous presence of aerodynamic, inertia and elastic (viscoplastic) forces must be considered. Under flight conditions, the complex physics associated with the presence of non-linearities in both the aerodynamics and structural / material behaviour, and fluid-structure interactions must be considered. To understand the consequences of this prediction, effective techniques for analysing the response, such as structural failure and aerothermoelasticity effects are required. The control aspect of aerothermoelasticity is a separate field and is beyond the intended scope of this chapter.

This chapter attempts to provide an overview of the state-of-the-art and insight into the key aerothermal issues / challenges for aerothermal and structural failure analysis of high-speed flight vehicles. The concerns of the aerospace vehicle designer and service life / failure analyst arising from structural integrity under the extreme aerothermal environments are placed in this context. The capabilities and problems of the currently available predictive methodologies, with a focus on those that can be readily implemented, are summarized and examples are presented to illustrate the approaches for aero-thermo-structural analysis under high-speed (high supersonic / hypersonic) flight test conditions. Lastly, some future directions are discussed.

2. PREDICTION METHODS

In this chapter, the predictive methodologies are organised into three major categories. These three groups are as follows: analytical techniques for 1) defining the complex loading environment (heating distributions and aerodynamic forces and moments), 2) predicting the thermal, structural and vibrational response and 3) predicting the consequences, such as structural failure and aerothermoelastic effects.

It is not the intention of this chapter to present an exhaustive review of all the available predictive methods; however the discussions are aimed at giving the reader an overview of the key issues and challenges and a feel for the type of computations used today. The reader is encouraged to consult the references to examine each topic in more details. Section 3 gives a detailed description of some numerical techniques that are practical and sufficiently accurate, and can be readily implemented as part of an initial design strategy or structural failure analysis.

2.1. Definition of the Complex Loading Environment

Aerodynamic loads exerted on the external surfaces of flight vehicles include pressure, skin friction and aerodynamic heating. Hence, characterizing the environmental loads is very important, especially for high-speed flights where the vehicle will experience increased aeroheating, aerodynamic, as well as, aero-acoustic loads. In high-speed flights, aerodynamic heating is the predominant structural load. Pressure and skin friction have a significant effect on the aerodynamic forces and moments, and therefore the flight stability of the vehicle. Additionally, acoustic fatigue can occur when fluctuating pressures produced by the vehicle and / or the external boundary layer cause the structure to vibrate at its natural frequencies. In particular, the aero-acoustic environment associated with hypersonic flight in the atmosphere is severe. There are several classes of hypersonic vehicles (for example, the National Aerospace Plane (NASP) and the newer hypersonic configurations such as NASA's X-33 or X-34 launch vehicles) that differ in their susceptibility to acoustic fatigue; in all cases, the aeroacoustic environment must be factored into the vehicle design. Thermo-acoustic fatigue issues are treated in Chapter 2 of this book. It should also be noted that modern high-speed aerospace vehicles are designed carefully within known parameters to avoid aeroelastic failure (i.e. to operate within the flutter envelope). Much work on aeroelasticity has been documented in the literature and a detailed review of this area and the control aspects are not the subjects of this chapter.

2.1.1. Aerothermodynamics

Aerodynamic heating is a critical design factor particularly for long duration flight vehicles operating at low altitudes, hypersonic systems and ground-to-air rocket systems with extremely high acceleration that operate at high Reynolds number (of the order 10^7 to 10^8, i.e. in the turbulent flow regime). The heating is a result of the high temperature gradient produced by the conversion of velocity into heat by viscous forces within the boundary layer that surrounds the vehicle. Hence, it is necessary to predict the flow over a high-speed body accurately in order to calculate properly the temperature in the shock layer and heat transfer to the surface of the body. An accurate description of the high-enthalpy shock layer flow field surrounding the body is also essential to predict the surface pressure distributions required for calculating aerodynamic coefficients (such as C_D, C_L, C_M), as well as the convective heat transfer rate required for the thermal-structural analysis. In depth discussions of aerothermodynamics - the characteristics of the flow over a high-speed body where the high-energy flow is governed by thermodynamics and mechanical laws - can be found in the textbooks by J.D. Anderson Jr. [1] and J.J. Bertin [2].

Theoretically, all of the equations for the different mechanisms of heat transfer (i.e. convective heating, radiation, and so on) must be solved simultaneously to determine the surface temperature history of the body. However, in practice only the heat transfer due to forced convection on the vehicle surface and then the dissipation of that energy through the structure is considered. Radiation is usually small, especially for applications inside the atmosphere, and it is assumed that the heat transfer is predominantly by convective heating.

The basic parameters in aerodynamic heating are:

1) heat transfer rate
2) heat transfer coefficient, h
3) recovery temperature T_r
4) Stanton Number, St
5) Reynolds Analogy Factor
6) reference temperature, T^*
7) Compressible to Incompressible Transformations.

Some of the many available methods and equations for calculating convective heat transfer rate and coefficient are described in more details in Section 3.2. The recovery temperature is important in heat transfer data correlation and extrapolation, in particular for applications where the difference between the recovery temperature and wall temperature (T_w) is small. It accounts for the part of the freestream energy that is converted to kinetic energy and the part that is dissipated as viscous work in a thermodynamically irreversible process and is evaluated using a recovery factor, r ($r = (T_{aw} - T_e)/(T_{st} - T_e)$, where T_e is the temperature at the edge of the boundary layer, T_{st} is the stagnation or total temperature and T_{aw} is the adiabatic wall temperature). The recovery factor is related to the Prandtl number by $r = Pr^n$, where $n=1/2$ for laminar flow and $n=1/3$ for turbulent flow. The Stanton number, St, is the ratio of actual heating rate to the total potential heating rate of the flow and is defined as $St = H/(\rho_e U_e C_p)$ where H is the enthalpy, ρ_e and U_e are the density and velocity of the flow at the edge of the boundary layer and C_p is the specific heat. The Reynolds Analogy Factor, S, relates heat transfer to skin friction and is given by $S = 2St/C_f$ where C_f is the skin friction coefficient. There are many formulations of the reference temperature, T^*, which is defined as a characteristic temperature at which the compressible friction factor agrees with the incompressible case. The Eckert formulation of the reference temperature is the most widely used and is given by: $T^* = 0.5(T_w + T_e) + 0.22r(T_{ste} - T_e)$, where T_{ste} is the total temperature at the edge of the boundary layer. The functions for transforming compressible (supersonic and hypersonic flow fields) to incompressible flow are functions of wall, stagnation and local temperatures. They enable the transformed skin friction and Reynolds number to satisfy incompressible skin-friction equations. There are a number of techniques for reducing a compressible flow heating problem to an incompressible flow problem and some are presented in references [6, 39].

Section 3 discusses some of the methods used in the prediction and correlation of aerodynamic heating. For flow over simple vehicle geometries at zero angle of attack, use of the basic engineering methods does not present difficulties. However, as the angle of attack (α) increases, pressure gradients are set up about the vehicle that causes streamlines to wrap about the vehicle, following regions of lower pressure (see Fig. **1**). Hence, the boundary layers in the regions of higher pressures are thinned, resulting in higher heating. In these cases and for vehicles with more complex geometries, Computational Fluid Dynamics (CFD) is used to provide a more accurate description of the inviscid and viscous flow fields.

2.1.2. Non-linear Aerodynamics

In general, the topic of aerodynamics of high-speed vehicles is understood relatively better than the thermodynamics. Only a brief discussion on this subject will be given here as it is well documented in the literature [41]. The discussion below is aimed at bringing out the pertinent material and data to enable the aerodynamic loads to be determined for structural design / failure / aerothermoelastic analysis.

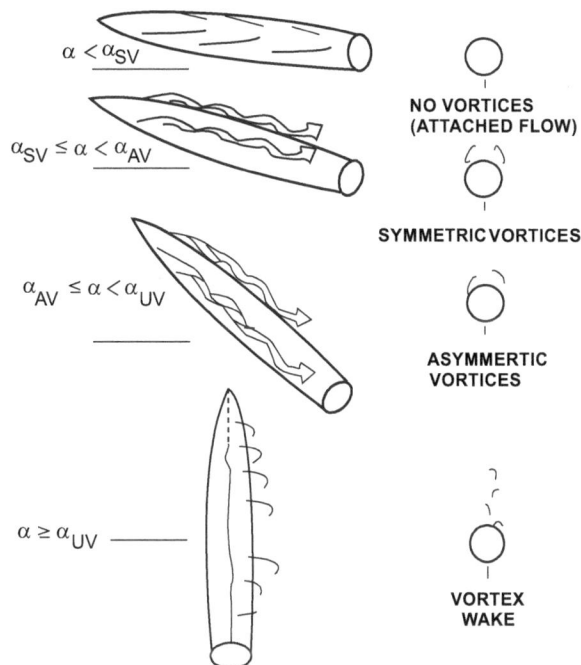

Figure 1: Effect of angle-of-attack on leeside of flow field, as taken from reference [5]. The subscripts SV, AV and UV denote symmetric vortices, asymmetric vortices and unsteady vortices respectively.

It is well known that non-linear aerodynamic behaviour becomes important at high supersonic and hypersonic speeds [3, 4]. The non-linearities arise from (1) the non-uniform inviscid flow field (entropy layer) generated by the curved bow shock (Fig. 2). This results in dynamic pressure gradients in control surfaces (fins, wings, tail, and so on), and is further complicated by the slight nose bluntness necessitated by thermal survival of the leading edge, (2) separated flows [5] in complex geometries (shock-induced separation and boundary layer separation on the surface of the body) and (3) viscous-inviscid interaction. It has been documented that viscous-inviscid interaction can cause large adverse effects on airfoil unsteady aerodynamics [9, 10] and severe aerodynamic heating. However, studies [11] using the first-order perturbation viscous method, extended to apply to the aerodynamic effects of elastic deformation of fins [11] indicated that viscous-inviscid interaction only has a small effect on the aerothermoelastic characteristics of a finned missile.

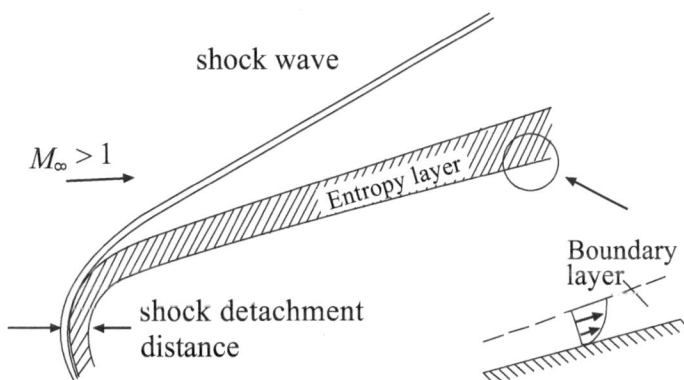

Figure 2: Entropy layer generated by bow shock.

Knowledge of both the inviscid and viscous flow fields over the vehicle is required for aerodynamic heating analysis and calculation of the generalized aerodynamic forces (GAF) needed in the aerothermoelastic analysis. As discussed earlier, the flow over a vehicle with simple geometry, for example a simple missile (wingless) shape at zero degrees angle of attack may be predicted using basic engineering methods [6], but as the angle of attack increases, pressure gradients developed about the vehicle that cause streamlines to wrap about the vehicle in regions of lower pressure resulting in thinner boundary layers in the high pressure regions, producing higher heating. Non-symmetrical structures and symmetrical structures at non-zero degrees angle of attack also have a pressure difference between their upper and lower surfaces which yield the GAFs that are responsible for the aeroelastic / aerothermoelastic response. Hence accurate prediction of the flow field about the vehicle is important for these cases.

For simple body geometries, the most common method used during the conceptual design of high-speed flight vehicles (such as tactical missiles) to obtain preliminary estimates of aerodynamic data (for flight performance and stability and control analyses) is the "component build-up method" [13, 52]. In this method the total loads for a vehicle is obtained by summing up the aerodynamic characteristics of the major part of the airframe separately and adding on the loads produced by component interference. This method has been widely used and produces rapid engineering data for conceptual design of subsonic and supersonic aircraft and missiles with simple geometries, where it is not necessary to account for the non-linear flow phenomena that are important in high supersonic and hypersonic flows.

Traditionally, the modelling approaches to unsteady aerodynamics in the high-supersonic / hypersonic regime were based on a number of approximate theories. The most commonly used methods to capture the unsteady components of the flow field are the Linear Piston Theory [64, 65] and Van Dyke's Second Order Piston Theory [66]. They compute the unsteady pressure distribution on the vehicle, due to its flexibility (rigid body motion) at hypersonic speeds and also fluid-structure interaction that arises when the vehicle vibrates. These theories are often used in studies on active control of aerothermoelastic effects [36, 67] as they allow the unsteady aerodynamic forces and moments (aerodynamic damping derivatives) to be included in the model of the flight dynamics of the hypersonic vehicle used in the control design process.

Other methods such as the Unsteady Shock Expansion Method [46] and Unsteady Newtonian-Impact theory [2], which assume inviscid high-speed flow, have often been used because they are sufficiently robust and can be readily implemented to give results suitable for conceptual design studies or for configuration parametric studies. These methods assume that the airflow over the vehicle is quasi-steady. Viscous effects are modeled with an effective shape or using the first order viscous perturbation model as reported in reference [53]. For example, the Newtonian Impact theory can be readily used to determine expressions for the pressures acting on the vehicle as a function of the Mach number, freestream pressure, angle of attack and vehicle geometry. These equations are then used to obtain the total aerodynamic forces acting on the vehicle and linearized to give the analytical expressions for the derivatives required for stability and control of the vehicle.

These relatively simple techniques for yielding pressure distributions / unsteady aerodynamics data are well documented elsewhere and it is not the intention here to give a detailed discussion of these methods. An example of application of the Unsteady Shock Expansion Method to calculate the pressure distribution on a generic hypersonic vehicle is described in Section 4 of this chapter.

All of these first principles methods can only provide a reasonable estimate of the pressure levels for flow conditions where the freestream conditions and the turning angles of the shock are known. For more accurate estimates of the pressure distribution, it is necessary to use CFD.

Recent advances in computational capabilities enable CFD solutions to the Euler and Navier Stokes equations to obtain the unsteady aerodynamic loads [14-17] and to understand the complex flow physics past complex configurations. The pressure variation across a thin (attached) boundary layer (laminar or turbulent), is negligible, so Euler solutions can be used as the surface pressure is given by the inviscid flow. Here the surface pressures are basically independent of the Reynolds number. However, when flow separation and viscous / inviscid interactions affect the pressure distributions, Navier-Stokes solutions must be used to model the viscous effects.

The Euler equations provide rapid and easily solved solutions for steady supersonic flows and predict aerodynamic coefficients, load distributions (forces and moments) and velocity profiles but viscous phenomena, such as body vortices could not be modeled. However, high-speed vehicles often operate in flow conditions where inviscid solutions are inappropriate. The Navier- Stokes solvers provide more realistic flight conditions which require the effects of viscosity to be included, but have large memory requirements and run times. Numerous comprehensive reviews and descriptions of the Euler and Navier-Stokes methods for supersonic and transonic flows can be found in the literature, for example, references [16, 17]. Reference [14] describes the use of the Euler/Navier-Stokes solver in the CFL3D code to perform both steady and unsteady flow calculations and aeroelastic analysis of a double-wedge airfoil. This code uses an implicit, finite volume algorithm based on upwind-biased spatial differencing to solve the time-dependent Euler and Reynold-averaged Navier-Stokes equations. The upwind-differencing is based on either flux-vector splitting or flux-difference splitting to enable sharp capturing of shock waves.

For high supersonic and hypersonic flows, the CFD solvers need to include viscous dependent heat transfer models in the analysis and this requires a more accurate temperature based treatment of boundary layer flow and transition, shock / boundary layer interaction, and so on. The CFD solvers must be able to provide good estimates of boundary layer transition, accurate shock capturing capability and turbulence performance. Automatic adaptive meshing with time-accurate evolution of the flow is desirable for investigating transient problems such as aerothermoelasticity and heat transfer under flight test conditions. Adaptive unstructured remeshing techniques have been developed to improve the efficiency and accuracy of steady-state and transient high-speed flow analyses by the finite element method, as well as structural analyses of the structure [54]. The size of the mesh changes itself both in time and space to capture accurately the transient responses. Further discussions on CFD analysis of high-speed flows will be given in Section 3.

2.2. Thermal-Structural-Vibrational Response

Aerothermal-structural analysis of high-speed vehicles requires 1) accurate generation and incorporation of realistic aerodynamic heating loads into the thermal-structural models and 2) transient coupled non-linear aero, thermal and structural analyses, to capture adequately the structural deformations and stresses in the vehicle encountered during rapid heating, for example during ascent or descent through the atmosphere [7].

The thermal response of the structure is coupled because the flow field properties (surface pressures and heating rates) are altered by surface temperature changing the amount of energy absorbed by the structure, and structural deformation of the structure resulting from the temperature gradients. Fluid-thermal-structure interaction has been demonstrated in a study [31] which shows Thermal Protection System (TPS) panels for re-entry vehicles thermally bending into the air-stream to relieve thermal induced stresses.

Furthermore, non-linear mechanics is needed to adequately predict the thermal-structural response associated with high-speed flights. The non-linearities come from the structure (arising from geometrical and/or material non-linearities) and also from the aerodynamics. As discussed earlier, hypersonic aerodynamics, in particular, can be highly non-linear, which arises from flow separation, shock wave interactions, and so on. Non-linearities in the structural mechanics must also be considered. The transient thermal loading conditions encountered in high-speed flights require the use of temperature dependent material properties (thermal conductivity, specific heat capacity, thermal expansion coefficient, and modulus) in the thermal-structural analysis. Additionally, the constitutive equations used in the structural analysis must be able to account for viscoplastic material behaviour, as well as the decrease in elastic moduli and strength with temperature [18, 28].

The changes in vibration characteristics also need to be considered for flexible structures and for missions where aerodynamic heating is considerable, such as when the vehicle is cruising at low altitudes or during pull-up maneuvers.

The operational requirements of some classes of hypersonic vehicles (such as the NASP, National Aerospace Plane, where the proposed flight envelope includes low subsonic to hypersonic speeds) cause the frequencies of these vehicles to be very low. It was reported in reference [36] that without aerothermal effects, the frequencies of a conceptual model of the NASP lie between 3 and 11 Hertz, but when the effects of heating were incorporated, the frequencies dropped by 13 to 20 percent, due to the de-stiffening effect of the aerodynamic load. This effect was also observed in another study of a generic hypersonic vehicle [7] where the modal frequencies dropped by 4 to 5 Hz when the trajectory changed from 0 to 30 km altitude. However, in this case the generic hypersonic vehicle behaved as a rigid structure – the lowest natural frequencies are relatively high and aerothermoelastic effects are not expected to be significant.

A flow diagram depicting a typical aerothermal-structural and vibration analysis is shown in Fig. **3**. Detailed descriptions of some of the methods for thermal-structural and vibrational analysis, and some example applications of these numerical techniques are given in Sections 3 and 4.

FLIGHT CONDITION
H(t), M(t), α(t)

↓

| CFD, basic engineering methods
Aeroheating flux,
surface temp, pressure distribution |

Temp dependent material properties
• constitutive models,
• thermal expansion coefficients, etc

↓

| Transient Heat Analysis | ←→ | Stress Analysis
Quasi-steady state
Mechanical Solver | → | Stress / strain
state of structure |

↓   ↓

| Structural modal analysis | | Failure Analysis |

↓   ↓

Modes and frequencies   Structural Integrity

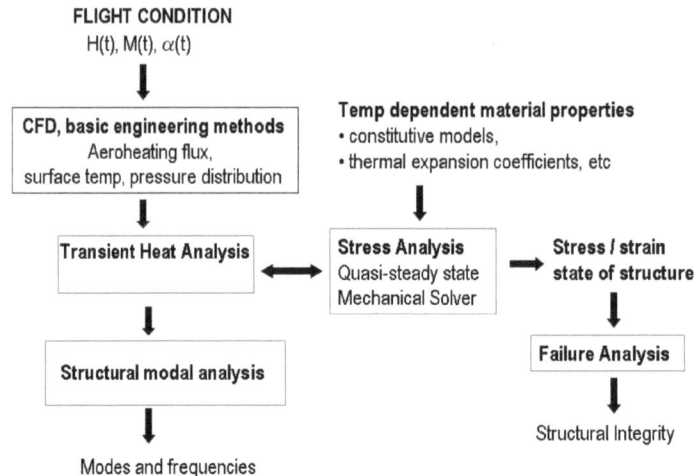

Figure 3: Flow diagram for aero-thermal-structural and vibration analysis, where *H(t)*, *M(t)* and *α(t)* are the altitude, Mach number and angle of attack at time, *t*, respectively.

2.3. Structural / Material Performance and Failure Criteria

An important step of the structural integrity analysis is to determine whether failure will occur and, if not, how far the given stress-strain state of the structure is from failure. This is particularly relevant for advance applications such as predicting the life expectancy of reuseable hypersonic vehicles which have been subjected to varying environmental conditions. Currently, there are two main methods to predict and evaluate failure: the deterministic structural analysis methods and probabilistic assessment methods.

The traditional deterministic approaches use allowables or margins of safety, computed based on selected failure criteria and knockdown factors (to account for variabilities) to obtain a conservative assessment of the viability of the structure. The margin of safety is a measure of the excess material capability over the design requirement. A value would need to be calculated at each critical location in the structure for various loading conditions, using an appropriate failure criterion. The selection of the failure criteria depends on various factors, such as loading condition (e.g. thermal, pressure, static, dynamic, combined loading, cyclic loading), failure type and material type (e.g. composites, metallic). The failure type can be described in several ways, including: 1) operational, such as deviation from engine performance, 2) failure related to the structure (including aerothermoelasticity effects), 3) corrosion resulting from chemical service environment.

This chapter is only concerned with failure type related to structural integrity. The failure modes include buckling, fatigue cracking, material yielding and so on. Structural deformation can also be considered a failure mode if the shape of the structure, for example distortion of the engine throat [7], is changed such that the performance of the vehicle is affected unacceptably. A widely accepted classification of failure criteria is according to the presence or lack of an initial flaw (crack). When there are no pre-existing flaws in the structure or if they are not large enough to influence fracture (crack propagation), the failure criteria used to predict failure include: maximum normal stress, maximum principal strain, maximum shear, maximum deviatoric stress, and so on. However, when the pre-existing flaws will influence fracture, a fracture mechanics or damage approach is required. In this case, the failure criteria used include strain energy release rate, stress intensity factor, damage, and so on. Advanced fracture mechanics models which could be used for predicting crack growth in aerospace structures are described in Chapter 5 of this book. Reference [58] provides an overview of the state-of-the art in the prediction of damage tolerance (defined as the ability of a structure to resist failure due to the presence of flaws, cracks or other damage for a specified period of usage). In general, failure is predicted when the predicted stress (or strain) at the critical location exceed the failure criteria. After the structural response under the given loads is determined analytically, a safety margin is computed based on the failure criterion and knockdown or safety factors. The safety margin must be a positive number for the structure to be structurally sound (i.e. $C/S \geq 1$ where C is the material's measured capability and S is the predicted stress or strain).

The probabilistic methods specify a level of reliability and are especially applicable to hypersonic structures where the environmental (aerothermal, acoustic, structural) loads are not well determined. In particular, reuseable hypersonic vehicles will be subjected to a variety of complex and severe cyclic loading conditions, including high heating rates and high temperature gradients (for example during ascent and descent trajectories). The reliability measure is the probability that the actual property random variable will be greater than the required property random variable. Reference [59] describes a methodology for probabilistic quantification of the uncertainties in the structural performance of aerospace hot structures, viz. the probabilistic load simulation, probabilistic finite element analysis, probabilistic simulation of thermomechanical nonlinear material behaviour and evaluation of reliability and risk. In that study, the uncertainties in all the basic parameters (variables) for loads, structure and material behaviour are incorporated to obtain a probabilistic simulation of the uncertainties in the structural response. It was noted that this methodology could readily be incorporated to monitor the in-service health of aerospace hot structures.

2.4. Aerothermoelastic Effects

Aerothermoelasticity of high-speed vehicles is a wide ranging and complex discipline. It can be defined by the structural deformation and /or dynamic response due to the mutual interaction of inertial, elastic and aerodynamic forces and heat input [12]. Studies have found that aerodynamic heating altered the aeroelastic stability of hypersonic vehicles (NASP) through the degradation of material properties and thermal induced stresses [19, 36]. Although the field of aircraft aeroelasticity has been widely published, the amount of published research on the aerothermoelasticity problem is very limited due to its complexity. The control aspect of aerothermoelasticity and aeroservoelastic issues, which considers the actuation and control system dynamics, are separate fields and are beyond the intended scope of this chapter.

At subsonic and low supersonic flight speeds, linear theories are used to describe the structural and aerodynamic behaviour in the aeroelastic analyses. The different stages of aeroelastic analyses (obtaining mathematical models of the structural modes and aerodynamic conditions for generating the unsteady aerodynamic responses) are treated separately and it is adequate to assume that the structural properties are invariant with time. Since the 1990s, there has been increasing effort in the development of aeroelastic analysis methods involving CFD techniques [20, 21]. A review of the methods for determining unsteady aerodynamics and aeroelasticity analysis will not be included in this chapter as a number of reviews on these computational methods can be found elsewhere [15, 37]. A general flow diagram for the computational aeroelastic and aerothermoelastic solutions is illustrated in Fig. **4**.

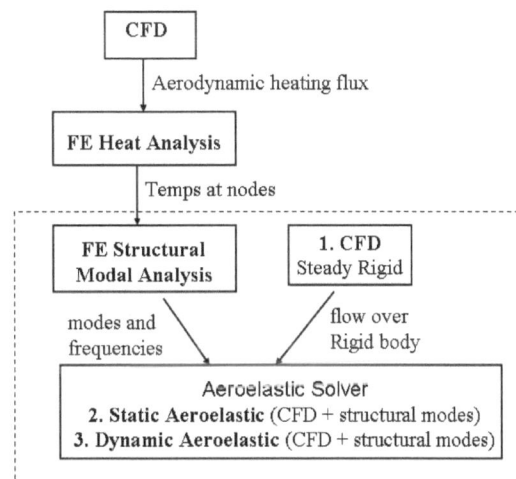

Figure 4: A general flow diagram for computational aeroelastic (dotted box) and aerothermoelastic solutions.

There is comparatively less work done in the aerothermoelasticity area. Early aerothermoelastic studies were based on linear theories to define both structural mechanics and aerodynamics, where the thermal, aerodynamic and structural effects were considered separately [12]. However, once the flight speed is in the high supersonic to hypersonic regime, nonlinear aerodynamics effects become important. The effect of non-uniform aerodynamic heating must be included and the analysis becomes much more complicated [9]. The aerothermoelastic problem is not steady-state and cannot be formulated in terms of the current flight condition. Both thermal and structural analyses must be transient and non-linear structural behaviour considered (material properties vary with temperature and geometric non-linearities). Aerothermoelastic analysis must consider all the couplings between aerodynamic, thermal and structural effects to account for the various nonlinear interactions.

A common approach is the loose coupling method to establish the aerothermoelastic model. A loose coupling of existing codes is achieved by selecting a reference coordinate system for a specific variable and mapping / interpolating that variable to the other codes at a given time step [7, 8, 32, 38]. The analysis consists of the following steps and computations: 1) aerodynamic heating, 2) temperature distribution in the structure, 3) thermal-vibration analysis, to determine the evolution of mode shapes and natural frequencies due to temperature changes and 4) aeroelastic analysis. As described in reference [42], the traditional method for generating aeroelastic response from CFD is very computationally intensive and usually involves the following steps: i) perform a converged steady rigid solution for a given Mach number and angle of attack to obtain the initial conditions to ii) perform a converged static aeroelastic solution at the same Mach number, angle of attack and a selected dynamic pressure and velocity to obtain the initial conditions to iii) perform a dynamic aeroelastic response at the same Mach number, angle of attack, dynamic pressure and velocity. This process must be repeated at several conditions to get stability / instability boundaries.

An integrated platform, based on a four-field formulation of aerothermoelasticity, for simulating the fluid-structure-thermal interaction was proposed by Tran and Farhat [34]. The aerothermoelasticity was formulated by the following four coupled equations: 1) the ALE (Arbitrary / Mixed Eulerian Lagrangian) non-dimensional conservation form of the Navier Stokes equation 2) the non-linear form of the structural equations of dynamic equilibrium 3) the governing equation for heat transfer in a structure 4) equation describing the motion of the fluid mesh assimilated to a quasi-static pseudo-structural system. A partitioned or staggered procedure was used to solve the system of coupled equations. In this work, only a one-way thermal mechanical coupling (i.e. a change in temperature causes deformations and stresses but not conversely) was considered. This solver was used to study the aerodynamic heating of an F-16 airfoil and a flat panel. Aerodynamic heating of the panel reduced the flutter boundary and increased the oscillation amplitude.

Recently, Silva [35, 42-43] presented a new approach using reduced-order models (ROM) for computing aeroelasticity. This approach has two significant advantages: firstly, the CFD results are in a form that could be utilized by other disciplines such as controls or optimization, and secondly, considerable reduction in the computational cost. In this approach, the ROM of the unsteady aerodynamic system, in state-space form, was developed using the CFD-based modal impulse responses. The unsteady aerodynamic state-space ROM was then combined with the state-space model of the structure to create an aeroelastic simulation using the MATLAB/SIMULINK environment [35]. Later improvement [42-43] to the ROMs approach involved the simultaneous excitation of the structural modes of the CFD-based unsteady aerodynamic system. This enabled the computation of the unsteady aerodynamic state-space model using a single CFD execution, independent of the number of structural modes. The method presented enabled the computation of matched-point (i.e. the corresponding atmospheric conditions for a given altitude and Mach number) solutions using a single ROM that is applicable over a range of dynamic pressures and velocities for a given Mach number. It should be noted, however, that the ROMs solutions currently do not include the simulation of aerothermoelasticity effects.

3. DESCRIPTION OF NUMERICAL TECHNIQUES

As can be appreciated from the above discussion, the aerothermal-structural performance of high-speed flight vehicles is characterized by multidisciplinary interactions of a number of technical disciplines, i.e., aerothermodynamics, fluid mechanics, structural mechanics and dynamics. This section provides a more detailed description of some of the many numerical techniques, which are applicable to high-speed flight vehicles, for predicting the 1) complex loading environment (heating distributions and aerodynamic forces and moments), 2) thermal, structural and vibrational response and 3) consequences, such as structural failure and aerothermoelastic effects. It is not intended to be an exhaustive list of all the very best techniques, as this is impossible in the space allowed here. Emphasis is given to those methods that can be readily implemented and are sufficiently accurate and robust for practical problems. The techniques used to generate the results in the example applications presented in Section 4 will be presented below. Alternative approaches will also be identified.

3.2. Trajectory Simulation

The aerodynamic and heating loads are determined by the flight conditions (altitude, angle of attack and Mach number) and vehicle geometry. As discussed earlier, the heat transfer in high-speed vehicles is transient and it is important to consider the operating conditions through the flight time. Hence, it is necessary to know the flight trajectory, in order to compute the transient heating and aerodynamic loads seen by the vehicle.

The main function of the trajectory analysis is to obtain an optimal trajectory that minimizes the fuel (total vehicle weight) while satisfying other constraints such as Mach number, final velocities, altitudes, launch angle, and so on. For the aerothermal-structural analysis, trajectory simulation is performed to estimate the flight path of the vehicle which is necessary for realistic determination of the heating loads for the given flight condition and providing the altitude vs. time data for the transient thermal-structural analysis.

The vehicle dynamics can be modeled by numerically integrating the two or three-DOF (degrees-of-freedom) equations of motion, where the translation and rotation of a body are decoupled. To model the vehicle dynamics more accurately, a high fidelity 6 DOF model is required. Six-DOF trajectory simulations can be performed using the MATLAB/SIMULINK environment. More details on this are discussed in Section 4.

3.3. Heat Flux Calculation

3.2.1. Approximate Engineering Methods

Although the best method for predicting aerodynamic heating is viscous CFD solutions, which provide a direct calculation of the heat flux and understanding of the interactions between the inviscid and viscous flow regions, it is computationally intensive. Furthermore, the accuracy of the empirical turbulent models used in the viscous CFD solutions for turbulent flows is somewhat questionable. Hence, basic engineering methods, where the vehicle is subdivided into tractable local flow regions, have been the way design was conducted for high-speed flight vehicles. This technique was applied to the Space Shuttle, as presented in Fig. 5 and taken from [26].

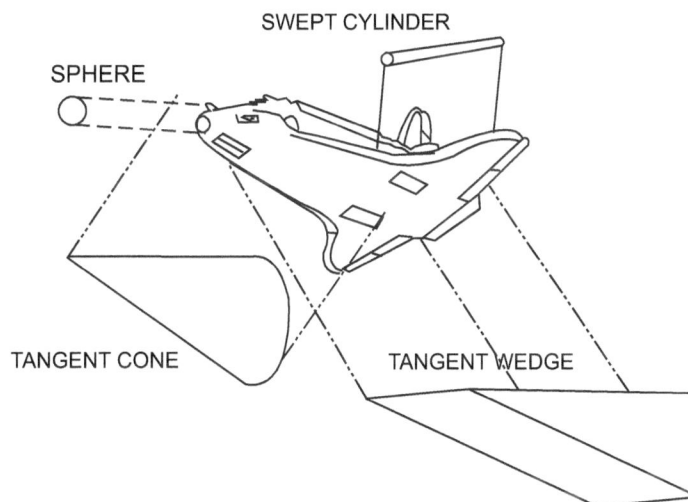

Figure 5: Analysis technique applied to the Space Shuttle, as taken from reference [26].

Reference [27] provides a method for calculating the time histories of transient surface temperatures and aerodynamic heat flux. This method was used to predict the transient surface temperatures and heating rates on the high-speed flight vehicles: X-15 research plane, YF-12 airplane, Space Shuttle, TU-144, Pegasus Hypersonic Experiment and Hyper-X. The results from this method showed good agreement with the measured flight data from the X-15 research vehicle, YF-12 airplane and Space Shuttle Orbiter.

The equations for predicting transient surface temperature and heating rates for stagnation points, and for laminar and turbulent values with transition for flat plates, wedges and cones, as described in detail in reference [27], are summarized below.

<u>***Stagnation Point***</u>

Heat flux (Btu/ft^2-sec) for 3-D stagnation points is:

$$\dot{q} = \left(\rho_w C_{p,w} \tau\right)\dot{T}_w = (F)(h)\left(H_{st} - H_w\right) - \beta T_w^4 + S \tag{1}$$

where ρ_w = density of wall material (lbm/ft^3), $C_{p,w}$ = specific heat of wall material (Btu/lbm°R), τ = wall or skin thickness (ft), $\dot{T}_w$ = rate of change of wall temperature (°R/sec), h = heat transfer coefficient (lbm/ft^{2}sec), F = empirical factor estimated from q/q_0, H_{st} and H_w = stagnation and wall enthalpy respectively (Btu/lbm), βT_w^4 = heat loss by radiation from surface of vehicle to atmosphere, S = solar radiation input if required (negligible except for low-speed flow).

The heat transfer coefficient (h) as given by Fay and Riddell (used for 3-D stagnation point) for Lewis number of 1 (no dissociation) and Prandtl number of 0.71 is:

$$h = 0.94\left(\rho_{st}\mu_{st}\right)^{0.4}\left(\rho_w\mu_w\right)^{0.1}\sqrt{\left(\frac{du}{dx}\right)_{x=0}} \tag{2}$$

and as given by Beckwith (used for 2-D stagnation point) with a Lewis number of 1.0 and Prandtl number of 0.71 is:

$$h = 0.704\left(\rho_{st}\mu_{st}\right)^{0.44}\left(\rho_w\mu_w\right)^{0.06}\sqrt{\left(\frac{du}{dx}\right)_{x=0}} \tag{3}$$

where the stagnation gradient (1/s) is given by:

$$\left(\frac{du}{dx}\right)_{x=0} = \frac{1}{R}\sqrt{\frac{2\left(P_{st} - P_1\right)g}{\rho_{st}}} \tag{4}$$

and the stagnation enthalpy for 3-D flow is calculated from:

$$H_{st} = H_2 + \frac{V_2}{2gJ} \tag{5}$$

where R = radius of nose or leading edge (ft), P_{st} = stagnation pressure (lb/ft^2), P_1 = static pressure in front of shock, g = gravitational conversion factor (32.17 lbm ft/lb sec^2), V_2, H_2 and P_2 = velocity, enthalpy and pressure behind the normal shock, J = mechanical equivalent of heat (778 ft lb/Btu).

<u>***Regions far from Stagnation Point (Constant Entropy)***</u>

The constant entropy solutions for the heat flux (Btu/ft^2-sec) at a flow distance x away from the stagnation point are given below. Reference [27] also gives a method for calculating heating for variable entropy flow.

$$\dot{q} = \left(\rho_w C_{p,w} \tau\right)\dot{T}_w = (h)\left(H_R - H_w\right) - \beta T_w^4 + S \tag{6}$$

where H_R and H_w = boundary layer recovery and wall enthalpy (Btu/lbm) respectively.

To solve equation (6), the heat transfer coefficients are calculated as follows:

<u>***Laminar heat Transfer***</u>

$$h = (F)\frac{0.332}{\sqrt{Re_L}}\sqrt{\frac{\rho^*\mu^*}{\rho_L\mu_L}}\left(Pr_w\right)^{-0.6}\left(\rho_L V_L\right)$$

$$= (F)0.332\sqrt{\frac{\rho^*\mu^* V_L}{x}}\left(Pr_w\right)^{-0.6} \tag{7}$$

where ρ_L, V_L are the density and velocity at the edge of the boundary layer, the superscript * refers to flow properties evaluated at the reference enthalpy given by:

$$H^* = 0.5\left(H_w + H_L\right) + 0.22\left(H_R + H_L\right)$$
$$where \quad H_R = H_L + \sqrt{Pr_w}\,\frac{V_L^2}{2gJ} \tag{8}$$

The values of H_w, H_L, T^* and μ^* are obtained from real gas tables.

Turbulent Heat Transfer

The heat transfer coefficient calculated by the van Driest method is given by:

$$h = F\,\frac{C_f \rho_L V_L}{2\left(Pr_w\right)^{0.4}} \tag{9}$$

Another method for calculating turbulent heat transfer coefficient uses the incompressible skin friction equation as follows:

$$h = F\,\frac{0.185}{\left(\log Re^*\right)^{2.584}}\left(Pr_w\right)^{-0.4}\left(\rho_L^* V_L\right) \tag{10}$$

The enthalpy H_w and Prandtl number Pr_w are functions of temperature and boundary layer edge static pressure P_L and are obtained from real gas tables. The transformational factors (F) in equations (7), (9) and (10) are 1.73 and 1.15 for laminar and turbulent flow respectively.

The constant entropy solutions are suitable for planar and conical bodies having attached flow on the leading edge. For shapes with a blunt surface, a detached shock is produced, with initially subsonic flow expanding to a supersonic flow field, i.e. blunting results in a flow field of variable entropy (due to curvature in the shock envelope) and pressures that deviate from the attached flow solution. The higher entropy regions also increase the local temperatures of the flow and correspondingly decrease the local velocity and Mach number. Variable entropy solutions for predicting the heating rates can be found in reference [27].

3.2.2. CFD Analysis

The time-dependent behaviour of a viscous, heat conducting laminar compressible fluid can be expressed by a set of partial differential equations governed by the conservation of mass, momentum and energy equations. The Navier-Stokes equations for 2-D flow in a Cartesian coordinate system can be written as:

$$\frac{\partial U}{\partial t} + \frac{\partial\left(E_I - E_V\right)}{\partial x} + \frac{\partial\left(F_I - F_V\right)}{\partial y} = 0 \tag{11}$$

where the solution (conservation variables) vector, U, and the flux vectors in the x and y directions, E and F, are given by:

$$U = \begin{Bmatrix} \rho \\ \rho u \\ \rho v \\ \rho e_t \end{Bmatrix}, E = E_I - E_V, F = F_I - F_V$$

The subscripts *I* and *V* denote the inviscid and viscous components of the flux vectors and are given by:

$$E_I = \left\{ \begin{array}{c} \rho u \\ \rho u^2 + p \\ \rho u v \\ \rho u e_t + p u \end{array} \right\}, E_V = \left\{ \begin{array}{c} 0 \\ \sigma_{xx} \\ \tau_{xy} \\ u\sigma_{xx} + v\tau_{xy} - \overset{\bullet}{q}_x \end{array} \right\},$$

$$F_I = \left\{ \begin{array}{c} \rho v \\ \rho u v \\ \rho v^2 + p \\ \rho u e_t + p v \end{array} \right\}, F_V = \left\{ \begin{array}{c} 0 \\ \tau_{xy} \\ \sigma_{yy} \\ u\tau_{xy} + v\sigma_{yy} - \overset{\bullet}{q}_y \end{array} \right\},$$

where ρ is the density, u and v are the velocity components, e_t is the total energy, σ_{xx}, σ_{yy}, τ_{xy} are the stress components and q_x, q_y are the heat fluxes. The compressible Euler equations are obtained by making the viscous flux vectors (E_V, F_V) equal to zero.

The pressure p in the inviscid flux components is defined by the state equation for a perfect gas as:

$$p = (\gamma - 1)\rho\left(e_T - \left(u^2 - v^2\right)/2\right) \tag{12}$$

where γ is the specific heat ratio.

The stress components, σ_{xx}, σ_{yy}, τ_{xy}, in the viscous terms are related to the velocity gradients by:

$$
\begin{aligned}
\sigma_{xx} &= \frac{2}{3}\mu(T)\left(2\frac{\partial u}{\partial x} - \frac{\partial v}{\partial y}\right) \\
\sigma_{yy} &= \frac{2}{3}\mu(T)\left(2\frac{\partial v}{\partial y} - \frac{\partial u}{\partial x}\right) \\
\tau_{xy} &= \mu(T)\left(\frac{\partial u}{\partial y} + \frac{\partial v}{\partial x}\right)
\end{aligned}
\tag{13}
$$

where $\mu(T)$ is the temperature dependent viscosity. The relationship for the absolute viscosity most frequently used in the literature is Sutherland's law:

$$\mu(T) = 1.46 \times 10^{-6}\left(\frac{T^{3/2}}{T + 111}\right) N.s/m^2 \tag{14}$$

The heat flux per unit area in the x and y directions are given by Fourier's law as:

$$
\begin{aligned}
\overset{\bullet}{q}_x &= -k\frac{\partial T}{\partial x} \\
\overset{\bullet}{q}_y &= -k\frac{\partial T}{\partial y}
\end{aligned}
\tag{15}
$$

where the thermal conductivity, k, can be calculated from the equation $P_r = c_p\mu/k$ using a Prandtl number, $Pr = 0.72$ and the specific heat at constant pressure, c_p.

The partial differential equations (11) must be discretised or transformed into a set of algebraic equations that can be solved numerically. Numerous methods are available to do the discretization, the three most popular are finite difference, finite volume and finite element. In the finite difference method, the partial derivatives are replaced with a series expansion, usually a Taylor series, typically truncated after 1 or 2 terms (more terms are included for more accurate solution but increases the complexity and number of nodes of the solution). In the finite volume method, the governing equations are integrated over a volume or cell assuming a piece-wise linear variation of the dependent variables. In these integrations, the fluxes (calculated at the mid-point between the nodes in the domain) across the boundaries of the individual volumes are balanced. In the finite element method, some form of Galerkin's method of weighted residuals, where the governing partial differential equations are integrated over an element after having been multiplied by a weight fraction, is generally used. The dependent variables are represented on the element by a shape function (for example, 2-D quadrilateral elements, 3-D hexahedral elements) which is the same form as the weight function. Other FE techniques employed include the Taylor-Galerkin or Petrov-Galerkin algorithms for high-speed compressible flow. Detailed descriptions of the solution algorithms for equation (11) using the finite element method can be found in references [23-25]. A number of commercial finite element based CFD codes are now available and they have successfully demonstrated the finite element discretization method in producing highly robust solutions of high speed turbulent flows as well as compressible flows.

3.3. Fluid/Structure Interaction

As mentioned in Section 2.2, significant interaction may occur between the aerodynamic flow field, structural heat transfer and structural response of a high-speed flight vehicle. Fluid-structure interaction has to be considered in the aerothermal-structural analysis in order to yield accurate predictions.

Finite difference and finite volume methods are the traditional and predominant numerical methods for CFD. A major difficulty in finite volume / difference CFD is the generation of realistic grids for realistic 3D geometries such as hypersonic vehicles. It is also difficult to perform fluid-structure interaction because the finite volume /difference based discretized model uses an Eulerian approach where the flow properties at a specific location in space over time are computed. Hence, accurate coupling of traditional CFD codes with finite element based structure codes is very difficult. On the other hand, the finite element model of the structure uses a Lagrangian approach where the properties at a point located on the structure (node) over time are computed.

The first simulation of fully coupled fluid-structure interaction, using a finite element (FE) based CFD code, has been reported by Thornton and co-workers [23-24, 32]. Efforts at NASA, university and industry to improve the capabilities and efficiency of finite element methods for high-speed inviscid compressible [22-23] and viscous flows [24-25] were aimed at enabling more efficient integration of fluid, thermal and structural analysis. Since the FE method can be utilized to discretize solids and fluids domains, its utilisation in modelling fluid-structure interaction ensures easier and more accurate coupling of the related disciplines. Also, the unstructured grids of FEM (finite element modelling) are more suitable for accurate discretization of complex geometries.

A loose coupling approach to the solution of fluid-structure interaction is reported in references [7, 8] where the aerodynamic heating loads from CFD are automatically interpolated onto the thermal- structural model using user-defined algorithms. Aerodynamic heating models as functions of spatial position and altitude are developed from the heating flux data for implementation in the thermal-structural analysis. An advantage of this approach is that it can be used to interpolate the aerodynamic heating data obtained from basic engineering methods [7], as well as CFD [8], onto the thermal-structural model. More details on this approach are discussed in Section 4.

3.4. FE Structural Heat Transfer

The Laplace equation, that governs the steady state flow of heat in a solid, can be used to describe the thermal response of the structure. For 3-D heat transfer this can be written as:

$$\frac{\partial}{\partial x}\left(k\frac{\partial T}{\partial x}\right)+\frac{\partial}{\partial y}\left(k\frac{\partial T}{\partial y}\right)+\frac{\partial}{\partial z}\left(k\frac{\partial T}{\partial z}\right)-\rho c_p\frac{\partial T}{\partial t}=0 \tag{16}$$

where ρ is the density, c_p is the specific heat capacity at constant pressure of the solid and $k\frac{\partial T}{\partial x}, k\frac{\partial T}{\partial y}, k\frac{\partial T}{\partial z}$ are the conduction heat fluxes in the x, y and z directions respectively.

The finite element form of representing the governing equation for transient heat transfer is:

$$K\{T\} + C\left\{\dot{T}\right\} = \{F\} \tag{17}$$

where K is the conductivity matrix, C is the damping matrix, $\{F\}$ is the applied load vector, $\{T\}$ is the vector of nodal temperatures and $\left\{\dot{T}\right\}$ is the rate of change of $\{T\}$ with time.

3.5. FE Structural Analysis

The structural response is governed by the quasi-static equations of motion. The 2-D structural response equations written in conservation form are:

$$\frac{\partial \{U\}}{\partial t} + \frac{\partial \{E\}}{\partial x} + \frac{\partial \{F\}}{\partial y} = 0 \tag{18}$$

where the displacement vector *{U}*, and the stress vectors components *{E}* and *{F}* are given by

$$\{U\} = \begin{Bmatrix} cu \\ cv \end{Bmatrix}, \{E\} = \begin{Bmatrix} -\sigma_{xx} \\ -\tau_{xy} \end{Bmatrix}, \{F\} = \begin{Bmatrix} -\tau_{xy} \\ -\sigma_{yy} \end{Bmatrix}$$

Here, c is a fictitious damping constant to assist marching to a steady-state quasi-static solution, u and v are the displacement components, σ_{xx} and σ_{yy} are the normal stress components and τ_{xy} is the shear stress component.

The finite element form of the linear dynamic equilibrium equation for calculating the transient response of a structure subjected to any arbitrary forcing function and initial conditions is as follows:

$$Kx(t) + C\dot{x}(t) + M\ddot{x}(t) = f(t) \tag{19}$$

where M is the mass (inertia) matrix, C is the viscous damping matrix, K is the stiffness matrix, $x(t)$ is the nodal displacement vector at time t, $\dot{x}(t)$ is the nodal velocity vector, $\ddot{x}(t)$ is the nodal acceleration vector and *f(t)* is the externally imposed forcing function (load vector).

For nonlinear behaviours, the above expression can be modified to include the nonlinearities, which are categorized into two main types: geometric and material nonlinearity. In a nonlinear solution, the stiffness matrix is continually updated to include the changes in geometry and stress redistribution within the structure due to the deformation and material yielding (in the case of material nonlinearity).

An example of geometric nonlinearity is a component or structure undergoing large deflections or deformations. Here, the solution is a function of the displacements or stress state of the structure and these are not known until the solution is completed. The solution must be obtained using load increments (rather than using each load level in one step) and an iterative algorithm used to obtain convergence of the unbalanced forces and increments of the displacement vector. The standard Newton-Raphson method is used in many commercial finite element structural codes.

Material nonlinearity must also be taken into account as the linear relationship between strain and stress is not adequate to model the material stress-strain relationship of inelastic components or structures subjected to high-speed flight, where the loading conditions require the use of temperature and strain-rate dependent material properties. The high temperature constitutive equations used in the structural analysis of high-speed aerospace vehicles are non-linear and must be able to take into account thermal damage, as well as the decrease in elastic modulus and strength with temperature. A method for developing constitutive equations that can describe the temperature dependence and non-linear behaviour of composite materials, and can be readily implemented in finite element structural analysis are described in references [18, 28]. At a given temperature, the constitutive equation has the form:

$$\sigma(T) = E'(T)\varepsilon^m + \eta(T)\dot{\varepsilon}^n \tag{20}$$

where σ is the stress, ε is the strain, E' is the Young's modulus of elasticity, η is the pseudo or artificial viscosity, $\dot{\varepsilon}$ is the strain-rate, m and n are the strain and strain-rate exponents respectively and T is the temperature.

Another constitutive model that has been used for finite element time-dependent thermoviscoplastic analysis of aerospace structures subjected to intense aerothermal loads is the Bodner-Bartom unified viscoplastic constitutive relation for rate-dependent non-linear material behaviour. Details of these equations and analyses are given in references [55-57]. In this thermoviscoplastic model, the strain rate-displacement relation is given by

$$\dot{\varepsilon}_{ij} = \dot{\varepsilon}_{ij}^E + \dot{\varepsilon}_{ij}^P = \frac{1}{2}\left(\dot{u}_{i,j} + \dot{u}_{j,i}\right) \tag{21}$$

where $\dot{\varepsilon}_{ij}$ is the total strain rate and the superscripts E and P denote the elastic and plastic components respectively. The derivative of the displacement components with time, differentiated over the spatial coordinates, are denoted by $\dot{u}_{i,j}$ and $\dot{u}_{j,i}$. The plastic strain rate is given by

$$\dot{\varepsilon}_{ij}^P = \frac{S_{ij}}{\sqrt{J_2}} D_0 \exp\left[-\frac{1}{2}\left(Z^2/3J_2\right)^n\right] \tag{22}$$

where D_0 is the limiting shear strain rate, n is strain rate sensitivity parameter (temperature dependent material parameter), S_{ij} is the deviatoric stress component, J_2 is the deviatoric stress invariant and Z is the internal state variable (can be interpreted as a load history dependent variable). Further details on evaluation of the plastic strain-rate can be found in reference [55].

As well as temperature dependent constitutive models, temperature dependent material properties, such as thermal expansion coefficient, specific heat capacity and thermal conductivity, in the temperature range appropriate to the flight conditions are also needed in order to accurately model the thermal-structural response of high-speed vehicles. These properties are not readily available in the literature. References [8, 29-30] provide some high temperature material data and describe some methods for obtaining those properties [29].

3.6. Structural Dynamic Analysis

As illustrated in Fig. **3**, the structural modal analysis is performed after the transient heat analysis to compute the natural frequencies and modes at a given trajectory point (altitude, Mach number, angle of attack). The temperature distributions from the transient heat analysis is applied to the structural dynamic analysis at different points in time over the duration of the flight, hence the changes in natural frequencies and mode shapes due to temperature change can be evaluated.

The FE structural model, equation (19), representing the vehicle equation of motion can be used to compute the natural frequencies (ω) and modes (ϕ). For free vibration, the applied force, *f(t)*, is set to zero. For an undamped structure, the damping matrix C is 0 and the natural frequency analysis is formulated as the following eigenvalue problem:

$$K\{\varphi\} = \omega^2 M\{\varphi\} \tag{23}$$

where K is the global stiffness matrix, M is the global mass matrix, $\{\phi\}$ is the vibration mode vector and ω is the natural frequency in radians/sec. Damping represents the loss of total system energy of the vibrating structure and can be modeled through the use of damping models or damping coefficients.

4. EXAMPLE APPLICATIONS

Aero-thermal-structural analysis of high-speed vehicles under flight conditions is a non-trivial undertaking as it encompasses many technical disciplines and requires the implementation and coupling of various multi-disciplinary analytical techniques. In this section, two numerical applications are presented to illustrate the various solution methodologies. The first concerns the aerodynamic heating and structural dynamics of a generic hypersonic engine under flight conditions, where the pressure distributions and heating loads were calculated using the oblique-shock theory / Prandtl-Myer flow equations, and approximate engineering methods [7]. The second concerns the aerodynamic heating of a research hypersonic vehicle with a complex 3D geometry, which results in complex flow phenomena around the vehicle, and the heating loads were calculated from CFD [8]. This example illustrates the use of aerothermal structural analysis as part of an initial structural design strategy.

4.1. Aero-Thermal-Structural Analysis of a Mach 10 Generic Hyshot Scramjet Engine

In this example, the aerodynamic heating of a Mach 10 generic HyShot [44] scramjet engine during re-entry into the atmosphere was studied. A typical HyShot mission profile is shown in Fig. **6a**. The scramjet (shrouded) is mounted inside the nosecone of a sounding rocket. The booster rockets propel the scramjet engine above the atmosphere. On re-entry to the atmosphere, an attitude control maneuver is performed on the vehicle to turn its nose pointing downwards. As the burnout rocket and payload fall back to earth, they reach a speed above M8 at around 35 km altitude when the scramjet experiment is started. The scramjet testing time is around 4-5 sec between 35 to 23 km altitude.

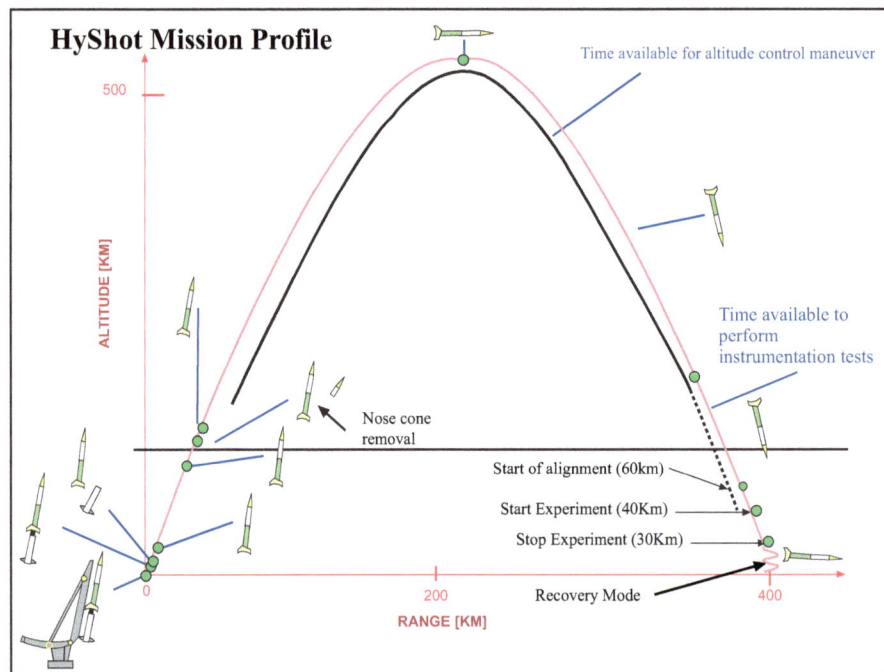

Figure 6a: HyShot mission profile, as taken from reference [44].

This generic HyShot scramjet has a simple 2-D geometry and it is assumed that the non-linearities in aerodynamics do not contribute significantly to convective heat transfer, which is a reasonable assumption because flow separation is unlikely to arise from the low angles of attack operation inherent in this type of scramjet design. The FE transient thermal-structural analysis was carried out on a 3-D model of the scramjet intake and combustor (see Fig. **7**), using a 3-D structured mesh with predominantly hexahedral brick elements. The scramjet components of interest are: an Aluminum alloy (6061) intake comprising three ramps with different wedge (turning) angles and a tungsten leading edge, the tungsten cowl at the entrance to the combustor, and a Copper alloy combustor. As the aim of this section is to illustrate the analysis of aerodynamic heating, the following discussion is limited to the aerothermodynamics and thermal-structural analysis of the intake ramps. Details on the aerothermo-structural analysis of the scramjet combustor can be found in references [7, 8].

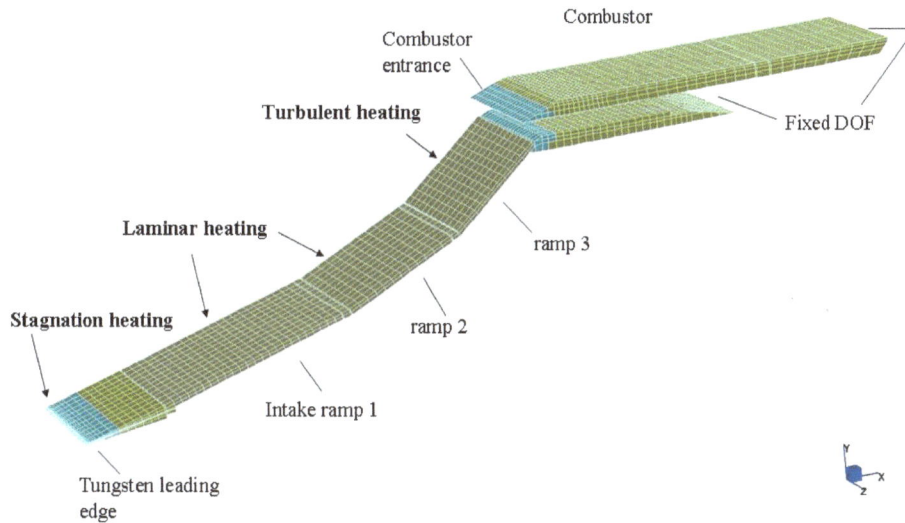

Figure 7: 3-D FE model of the combustor and intake of the M10 generic HyShot 2-D engine, as taken from reference [7]. Note: sidewalls not shown.

 The descent trajectory from 60 to 26 km was considered for the thermal-structural analysis, as this is the altitude range where the scramjet is expected to operate and is also the region that is most critical for the structural design of the engine as the effect of dynamic pressures on the heating loads start to become significant at altitudes below 40 km. The flight trajectory of the vehicle provided the altitude vs. time data for the transient heat analysis.

Trajectory Analysis

The nominal mission profile of this vehicle is similar to the HyShot 2 trajectory, as described in reference [44] (see Fig. **6a**). The flight performance of the vehicle can be calculated using a 6-degree-of-freedom trajectory simulation program or with the Aerospace Blockset in MATLAB/SIMULINK, as shown in Fig. **6b** [48]. The trajectory computed from the 6-DOF SIMULINK model compares well with the HyShot 2 trajectory calculated from the DTI proprietary 6-degree-of-freedom trajectory simulation program (see Fig.6c).

The trajectory simulation provided the altitude vs. time data for the transient thermal-structural analysis (for example, see Fig. **15**).

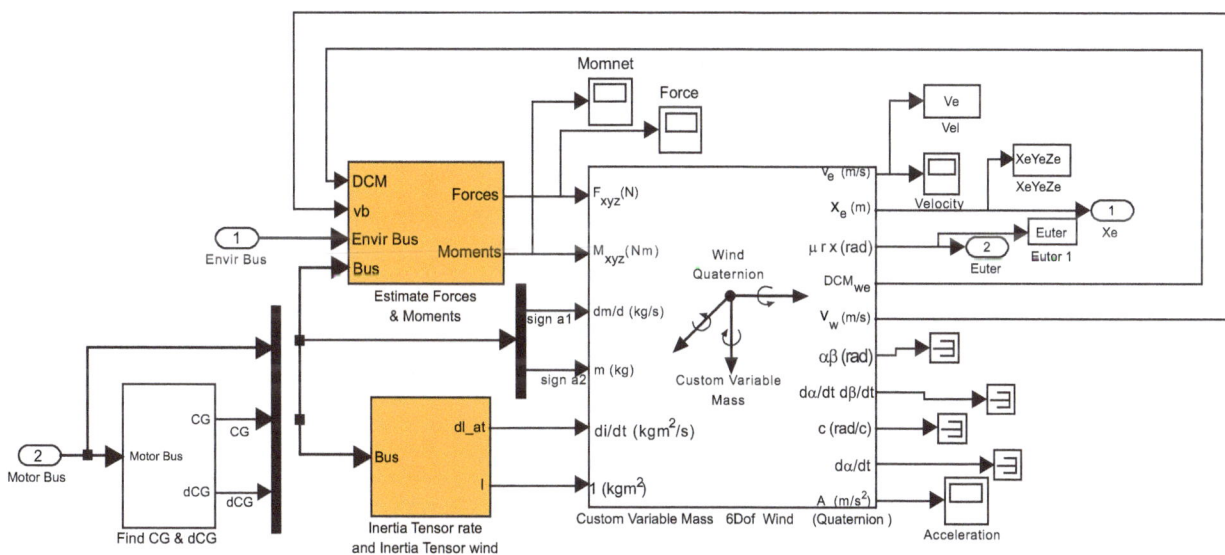

Figure 6b: Example of 6-DOF SIMULINK model for trajectory analysis of HyShot 2, as taken from reference [48].

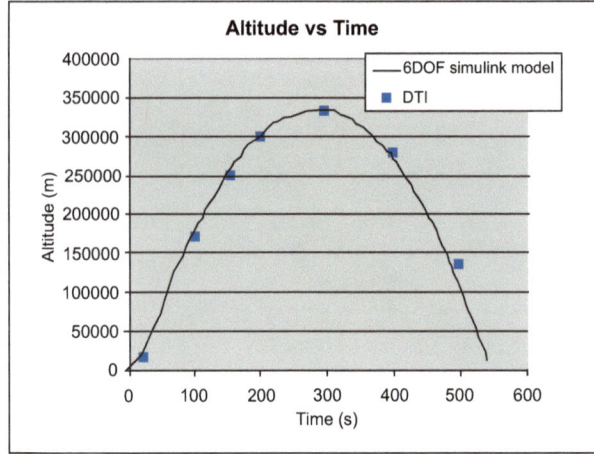

Figure 6c: HyShot 2 trajectory – comparison between simulations from 6-DOF SIMULINK model and DTI proprietary code, as taken from reference [48]

Prediction of Aerodynamic Heating Loads using Approximate Engineering Methods

In this study [7], the US Standard Atmosphere was used to represent the atmospheric density and temperature. The hypersonic approach flow over the intake and into the entrance of the combustor was calculated for various trajectory points from 60 to 26 km, using the standard compressible flow, normal shock and two-dimensional oblique shock functions. For example, Table **1** gives the local flow conditions in the stagnation region behind the normal shock on the intake leading edge and downstream of a series of oblique shocks for the intake ramps and cowl at 30 km altitude.

Table 1: Flow conditions in intake ramps for M10 flight at 30 km altitude.

Location	Plate Angle (deg)	Mach No.	Density (kg/m³)	Pressure (kPa)	Velocity (m/s)	Temp (K)
Freestream		10	0.0183	1.197	3022	226
Ramp 1	4.8	8.36	0.0379	3.479	2992	318
Ramp 2	9.3	7.27	0.0689	8.189	2959	412
Ramp3	14.69	6.29	0.1281	19.896	2914	539
Combustor Entrance		4.37	0.4183	122.2	2729	1013

The heat flux to a flat plate (intake ramps and entrance to combustor), assuming zero-pressure gradient, constant wall temperature and flat plate flows [3], is given by:

$$\dot{q}_w = \frac{0.332 p_e V_e \left(H_{aw} - H_w \right)}{RT^* \ Pr^{*\frac{2}{3}} Re_x^{*\frac{1}{2}}} \ \text{Laminar} \tag{24a}$$

$$\dot{q}_w = \frac{0.0287 p_e V_e \left(H_{aw} - H_w \right)}{RT^* \ Pr^{*\frac{2}{5}} Re_x^{*\frac{1}{5}}} \ \text{Turbulent} \tag{24b}$$

where, $Re_x^* = \dfrac{p_e V_e x}{RT^* \mu^*}$ (Reynold's number at reference temperature T*),

$Pr^* = \dfrac{\mu^* C_p^*}{\kappa^*}$ (Prandtl number at T*)

and p, V, T, H, μ, C_p, κ, x are the static pressure, velocity, temperature, enthalpy, viscosity, specific heat capacity at constant pressure, thermal conductivity and distance from the leading edge respectively. Subscripts e and w refer to edge or freestream and wall conditions respectively. The adiabatic wall enthalpy or recovery enthalpy, H_{aw}, represents the enthalpy of a perfectly insulated wall.

The reference enthalpy is given by:

$$H^* = \frac{H_e + H_w}{2} + 0.22r\frac{V_e^2}{2}$$

(25)

where, the recovery factor $r = \sqrt{\dfrac{\mu^* C_p^*}{k^*}}$ (Laminar)

and $r = \sqrt[3]{\dfrac{\mu^* C_p^*}{k^*}}$ (Turbulent)

The heat flux for stagnation point heating of a cylinder (leading edge with a small nose radius to reduce heating), for frozen boundary layer with a non-catalytic wall, is given by [1]:

$$\dot{q}_w = 0.57 \, \mathrm{Pr}^{-0.6} \left(\rho_e \mu_e\right)^{1/2} \sqrt{\frac{du_e}{dx}} \left(h_{aw} - h_w\right)$$

(26)

where, the stagnation point velocity gradient is given by Newtonian theory as:

$$\left(\frac{du_e}{dx}\right)_{st} = \frac{1}{R}\sqrt{\frac{2\left(p_e - p_\infty\right)}{\rho_e}}$$

(27)

and R is the nose radius. At the stagnation point, p_e and ρ_e are replaced by p_{st} and ρ_{st} respectively. The heat flux on a cylinder away from the stagnation point is then determined using a Newtonian pressure distribution from

$$p_e = \left(p_{st} - p_\infty\right)\cos^2\theta + p_\infty$$

(28)

where, subscripts st and ∞ refer to stagnation point and freestream conditions respectively and θ is the angle from the stagnation point.

To avoid discontinuities in the heat fluxes at corners of different geometries, the boundary layer origin of the flat plate solution was extrapolated so that the heat fluxes where the cylinder nose and flat plate meet were continuous.

The initial wall temperature was assumed to be 300 K for the heat flux calculations using equations 24-28. It was assumed that the flow over the first and second ramps is laminar and fully transitions to turbulent flow in the third ramp. CFD analysis [45] to determine the effects of turbulence within the boundary layers of this simple two-dimensional scramjet geometry, and the results presented in Fig. **8** show that this is a reasonable assumption. The results of the aerodynamic heating fluxes at various positions along the intake ramps as a function of altitude clearly show that turbulent flow does not have a significant effect on the convective heat flux above 40 km. Although the analysis covered a wide range of altitudes (60 km to 26 km), the heating loads do not become significant until below 35 km (where the dynamic pressure becomes significant) and hence, it is reasonable to assume a constant transition point. Available experimental and analytical evidence suggest that the accuracy of the convective heat transfer estimates based on the Eckert method is likely to be within ± 10-20% [46].

Models of the aerodynamic heating flux as functions of spatial position and altitude were developed from these results for incorporation into the FE thermal-structural analysis.

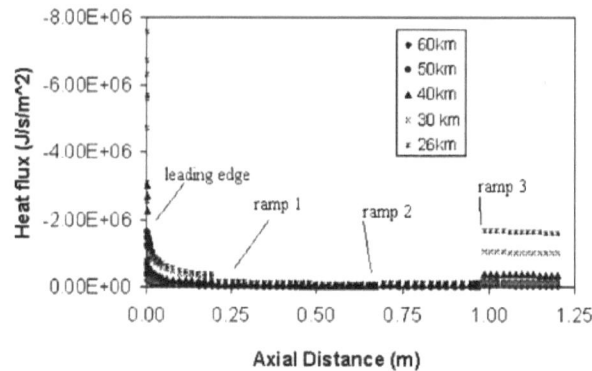

Figure 8: Heat flux as a function of axial distance and altitude, as taken from reference [7].

Transient Thermal-Structural Analysis

The heat fluxes from the aerodynamic heating models for varying flight conditions were automatically interpolated onto the 3-D FE thermal-structural models for each time-step in the transient thermal analysis using user defined algorithms in the FE code. More details on the development of the capabilities of this FE code can be found in reference [28]. The thermal loading was applied as a time-varying heat flux condition to model the change in environment at various altitudes in the flight trajectory. Using the calculated temperature-time history, a non-linear transient dynamic analysis is conducted. To accurately predict the stresses and strains induced by the thermal transient simulating the flight trajectory, temperature dependent non-linear material constitutive models were developed [18] and implemented in the transient dynamics solver of the finite element program.

The thermal transient in the thermal loading was defined by the flight trajectory. For this analysis, the start of the transient began at 60 km and ended at 26 km altitude (at time = 11.27 s). Boundary conditions for heat conduction to the sidewalls and radiation to ambient temperature were applied to the outside surface of the FE model. The transient dynamics solution assumes that the scramjet engine begins in a state of zero stress and strain at the start of the thermal loading at zero time. To determine the significance of the varying dynamic pressure with altitude on the thermal-structural response, the combined effect of thermal and pressure loading was conducted in the FEM. More details on this study can be found in reference [7].

The temperature contours in the intake ramps, with a plate thickness of 15 mm and nose radius of 2 mm, at the start and end of the thermal transient are shown in Fig. **9a-b**. Plots of temperature as a function of time (altitude) for the leading edge and the third ramp are shown in Fig. **10a**. As expected, the highest temperature (ca. 2008K) is at the leading edge, due to stagnation point heating. The temperatures increase exponentially with decreasing altitude. The difference between laminar and turbulent convective heat transfer is quite substantial at the end of the thermal transient – the temperature in the third ramp is 512K, approximately 190K higher than in the second ramp. Radiative heat transfer from the body surface to ambient temperature was found to have little effect in the relatively short time-scale of this analysis.

The plate thickness of the intake ramps, nose radius of the leading edge and length of the tungsten tip in the first ramp were varied in the analysis to determine the minimum length and thickness needed to maintain structural integrity of the scramjet. For the model with 1mm nose radius (compare Figs. **10b-c**), reducing the plate thickness from 15 to 10 mm resulted in a temperature increase of 65K for the third ramp and approximately 200K for the leading edge at time = 11.27 s (26 km altitude). Decreasing the nose radius from 2 to 1 mm (compare Fig. **10a** with Fig. **10b**) increased the temperature in the leading edge by approximately 132K (15-mm plate model). Also, reducing the length of the tungsten tip had little effect on the temperature in the first ramp for plate lengths greater than 70 mm, because the poor thermal conductivity of tungsten does not allow it to dissipate the heat. Although the predicted maximum temperature in the third ramp was well below the melting point of the 6061 Aluminum alloy (885 K), it is not clear from thermal analysis alone whether a minimum plate thickness of 10-mm may be adequate to withstand the thermal loads.

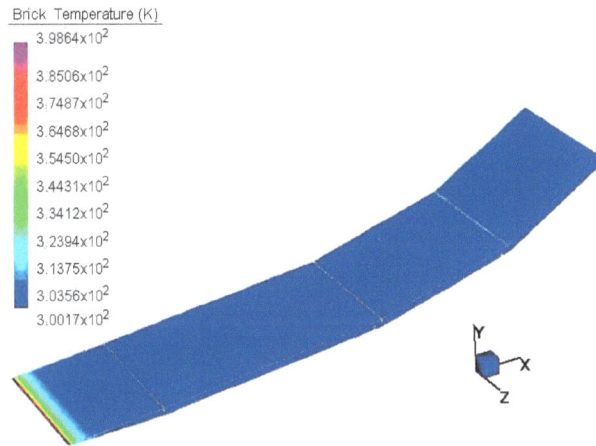

Figure 9a: Temperature distribution in intake ramps at the start of the thermal transient (60 km altitude), as taken from reference [7].

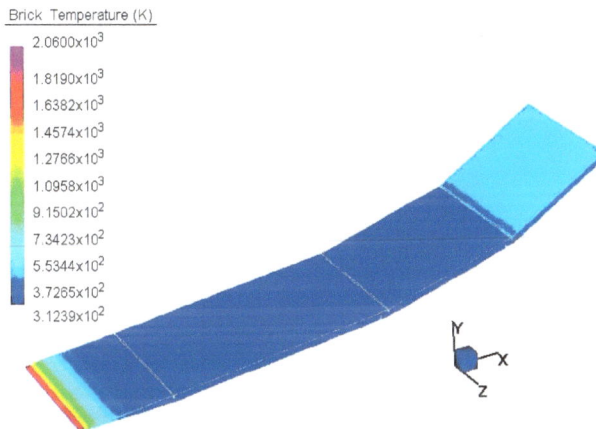

Figure 9b: Temperature distribution in intake ramps at the end of the thermal transient (26 km altitude), as taken from reference [7].

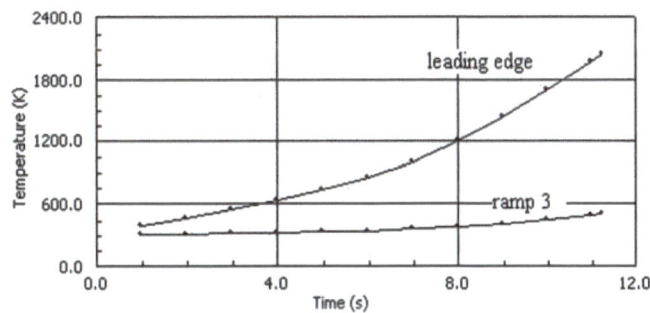

Figure 10a: Temperature as a function of time for the leading edge and third ramp (15-mm plate, 2-mm nose radius), as taken from reference [7].

To assess the likelihood of engine geometry distortion, the resulting thermal loads from the thermal analysis were used in the transient dynamic analysis to predict the deformation and stress-strain state of the structure during the thermal transient. Temperature-dependent material properties [49] (Young's modulus, thermal expansion coefficient, thermal conductivity, and specific heat capacity) were used where available. The deformed shape of the displacement contours in the intake at 30 km altitude (time ≈ 10 s) is shown in Fig. **11**. It shows bulging in the third ramp and to a lesser extent in the second ramp, in a manner consistent with a high temperature gradient between the upper and lower surfaces of the plate and constraint by the sidewalls. The maximum vertical displacement (DY) in the third ramp at 30 km altitude (time ≈ 10 s) is around 0.5 mm (10-mm plate model). The displacement contours also show the leading edge moving forward and the third ramp slipping towards the 3-mm gap, designed for stress relief, between the second and third ramps. The maximum axial displacement (DX) at 30 km altitude is 0.8 mm, in the leading edge (10-mm plate model).

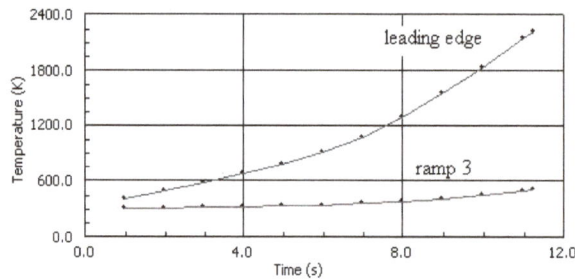

Figure 10b: Temperature vs. time plot for FE model with 15-mm plate and 1-mm nose radius, as taken from reference [7].

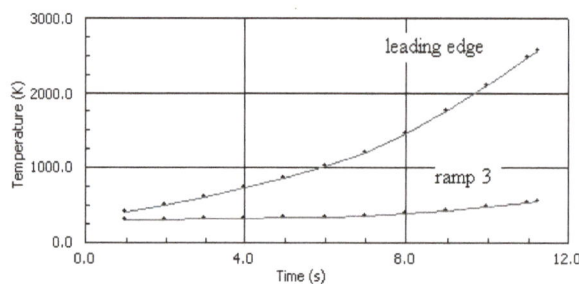

Figure 10c: Temperature vs. time plot for FE model with 10-mm plate thickness and 1-mm nose radius, as taken from reference [7].

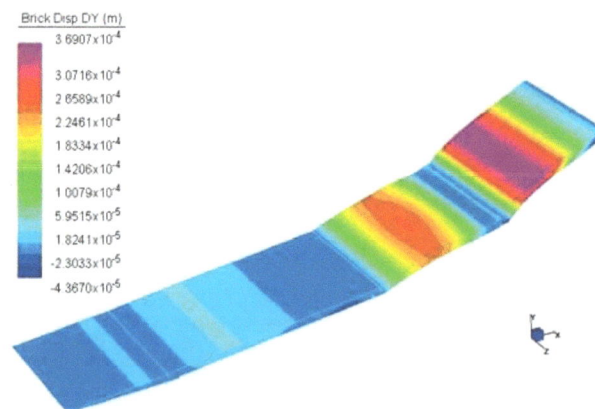

Figure 11: Displacement contours from transient thermal analysis at time = 10s (corresponding to 30 km altitude) for FE model with 15-mm plate and 1-mm nose radius, as taken from reference [7]. Displacement scale = 1%.

Von Mises stress and strain contours in the intake ramps at 30 km altitude (time ≈ 10 s) are shown in Fig. **12a** and b. The transient thermal loads induce stresses and strains in the scramjet mainly by the large temperature gradients and, to a lesser extent, the difference in thermal expansion coefficients between the leading edge material and the intake ramps material. The structural analysis predicts that the highest stresses are in the tungsten leading edge, as expected. For the 10-mm plate model, the predicted stresses in the second and third ramps are in the range 40 to 250 MPa. Our previous studies [28] suggest that the error in the stress analysis is likely to be ±20-30 %. The thermal induced stresses were well above the yield stress of 6061 Aluminum alloy at 600K (45 MPa), suggesting that the plate thickness should be increased.

The results indicated that deformation of the structure was not significant until at the end of the thermal transient. Hence the effect of the deformed structure altering the flow properties, and therefore the heating rate, was not considered here. This is a reasonable approximation since the heating loads vary slowly with time compared to the forces and displacements of the structure.

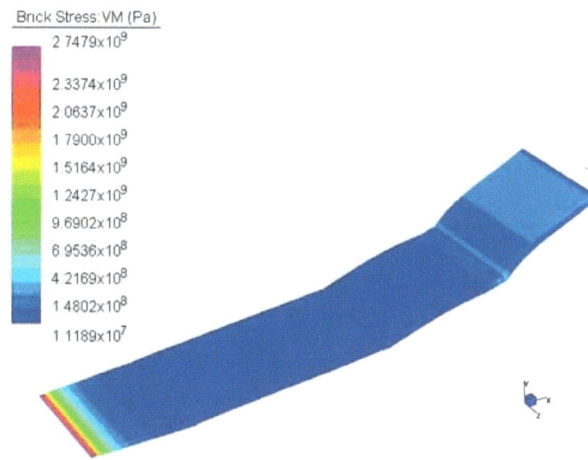

Figure 12a: von Mises stress contours in intake ramps at time = 10 s (30 km altitude) for FE model with 15-mm plate and 1-mm nose radius, as taken from reference [7]. Displacement scale = 1%.

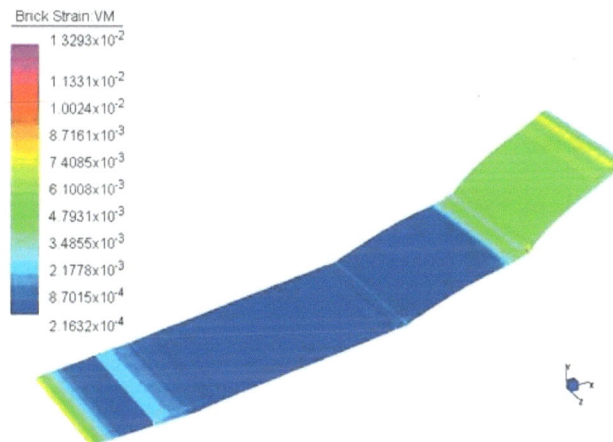

Figure 12b: von Mises strain contours in intake ramps at time = 10 s (30 km altitude) for FE model with 15-mm plate and 1-mm nose radius, as taken from reference [7]. Displacement scale = 1%.

Structural Vibration Modes Analysis

The natural vibration modes of the scramjet engine may be excited by flow disturbances (aerodynamic and combustion) and it is important to determine the changes in vibration modes and frequencies associated with thermal effects.

A vibration analysis was performed following the thermal-structural analysis, using the temperature distributions and resulting stress-strain state at any given time during the thermal transient, to study the effect of temperature changes associated with the descent trajectory from 60 to 26 km on the vibration characteristics of the structure. The nodal temperatures at a given time in the thermal transient are automatically mapped onto the FE model for the linear static and natural frequency analyses.

The modal shapes and natural frequencies of the model of the intake (10-mm plate) and combustor together, at a uniform temperature of 300 K, are shown in Fig. **13**. The lowest natural frequency of this model is 118 Hz, attributed to bending of the upper plate of the combustor. The mode 3 deformation is at 306 Hz and corresponds to bending of the first intake ramp.

The vibration characteristics of the intake ramps (10-mm plate) at 30 km altitude are compared with those at a uniform ambient temperature (300 K) in Table **2**. The mode 3 natural frequency, corresponding to bending of the first intake ramp, is 306 Hz at 300 K and shifted to 302 Hz at 30 km altitude. These results indicate that the lowest natural frequency for this type of scramjet engine for the descent trajectory studied here (see Fig. **13**) is high and is outside the frequency range where control issues are relevant. However, for operation at low altitudes over a longer period of time (for example, during a pull-up maneuver or cruising), aerothermoelastic effects may become more significant for this hypersonic vehicle.

Mode 1   118 Hz

Mode 2   241 Hz

Mode 3   306 Hz

Figure 13: Vibration modes and frequencies of combustor and intake ramps (10-mm plate) at a uniform temperature of 300K, as taken from reference [7]. Displacement scale = 1.5%.

Table 2: Vibration frequencies of intake ramps (10-mm plate) at 300K and 30 km altitude.

Mode	Description	Frequency (Hz)	
		300K (cold)	30 km (hot)
3	1st ramp bending	306	302
4	1st ramp torsion	505	500

4.2. Aero-Thermal-Structural Analysis of a Hypersonic Vehicle For Flight Test

The second example is a real-life application. The aerothermal-structral analysis was performed on the DARPA HyCause 3-D engine [50]. This research hypersonic vehicle has an 'inward turning' engine [51]. The complex 3-D geometry of this vehicle results in flow phenomena that are difficult to analyze using approximate engineering methods. Shock tunnel experiments performed at CUBRC and CFD analysis of this vehicle [51] showed regions of shock wave / turbulent boundary layer interaction resulting in aero-thermal loads that was of major concern to the structural design. The shock wave / turbulent boundary layer interaction inside the isolator and combustor of the scramjet engine dominated the flow structure, and the resulting large aerothermal loads must be taken into account in the structural design of the flight vehicle. As part of the design strategy, the heat flux (aerodynamic heating and combustion) data from CFD and shock tunnel experiments [51] were utilized to conduct a transient coupled thermal-structural analysis of the vehicle for the HyCause descent trajectory [8]. Only the aerodynamic heating aspect of the study is discussed here. Details on the thermal-structural analysis of the combustor can be found in reference [8]. The HyCause payload is illustrated in Fig. **14**.

Aerothermal-Structural Analysis of the Preliminary Design Model

The finite element transient heat analysis was conducted for a Mach 10 HyShot type flight trajectory during re-entry into the atmosphere. The descent trajectory from 60 to 37 km was considered here, as this is the altitude range where the HyCause engine is expected to operate. This is also the region that is most critical for structural design of the engine as the effects of dynamic pressures on the heating loads start to become significant at altitudes below 40 km. For this analysis, the start of the thermal transient began at 60 km and ended at 37 km at (time = 7.6 s), as shown in Fig. **15**. Turbulent heating loads and an initial cold wall temperature of 300K were used to obtain a conservative analysis for the structural design.

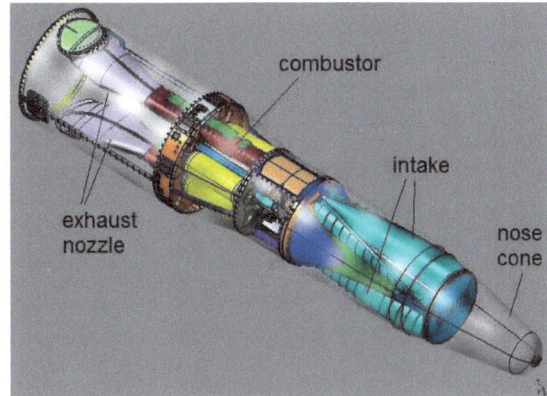

Figure 14: Drawing of HyCause Payload, as taken from reference [40].

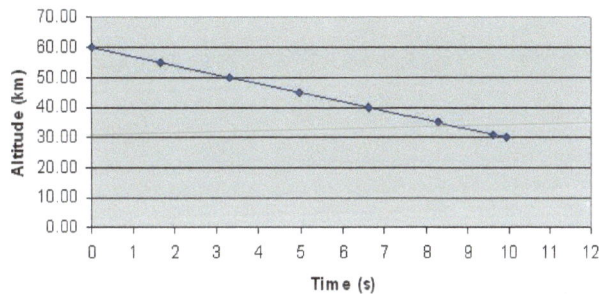

Figure 15: Plot of altitude vs. time for transient heat analysis.

The heat flux data were obtained from CFD [8, 51] and aerodynamic heating models as functions of spatial position and altitude were developed for interpolation onto the 3-D thermal-structural models for each time-step in the transient thermal analysis, using user-defined algorithms. The thermal loading was applied as a time varying heat flux condition to model the change in environment at various altitudes in the flight trajectory. Using the calculated temperature-time history from the transient heat analysis, a non-linear transient dynamic analysis was conducted to predict the thermal induced stresses and deformations in the vehicle during the thermal transient. Temperature dependent non-linear material constitutive models were developed from experimental data [29] and implemented in the FE program [28]. The transient dynamics solution assumes that the scramjet engine begins in a state of zero stress and strain at the start of the thermal loading at zero time.

As in the previous example, the effect of the deformed structure altering the flow properties and therefore, the heating rate was not considered in the thermal-structural analysis. This is a reasonable approximation since the heating loads vary slowly with time compared to the forces and displacements of the structure. Also, the results from this study showed that the deformation of the structure was not significant until towards the end of the thermal transient.

A 3-D FE mesh, with predominantly hexahedral brick elements, of the initial design (Preliminary Design Review, PDR, model) and the applied heating loads at various parts of the engine are shown in Fig. **16**. This model does not include the insert (tungsten or steel) in the crotch (for heat mitigation), injector block and expansion slots (for stress relief and weight reduction) on the leeward side of the intake. Bulkhead constraints were modelled using boundary conditions on the intake (at 930 mm) aft of the nose. The initial thermal-structural analysis modelled the entire engine structure, except for the leading edge, with Aluminum alloy T6 6061. The leading edge was modelled with different materials including tungsten, low carbon steel, annealed molybdenum, Copper alloy and Nickel alloy, to determine the most suitable material. The isolator and combustor have variable wall thickness as shown in Fig. **17**. The scramjet components that are of interest to the aero-thermal-structural analysis are the intake, leading edge, isolator and combustor.

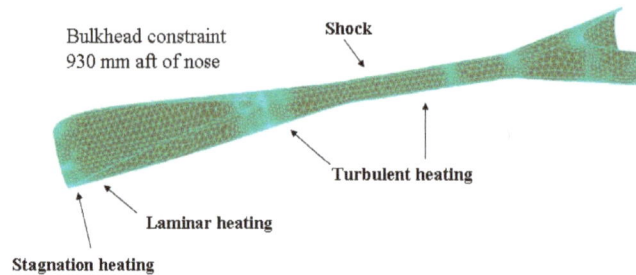

Figure 16: 3-D finite element model of the PDR engine and applied heating loads, as taken from reference [8].

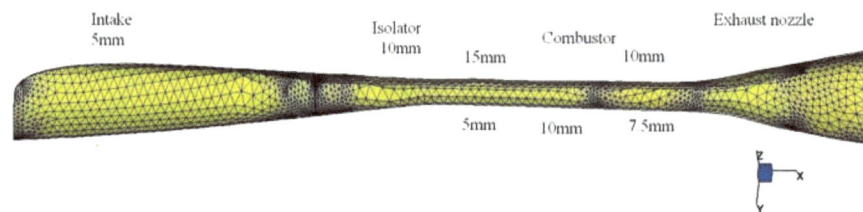

Figure 17: 3-D finite element half model of PDR engine, as taken from reference [8].

The temperature distribution at the end of the thermal transient (37km altitude) from the transient heat analysis of the PDR model for the fuel-off condition is shown in Fig. **18**. The maximum temperature at 37km altitude is around 700K in the isolator and in the thin sections of the combustor duct. The temperatures in the intake (5mm wall thickness) vary from around 1200K at the leading edge (depending on the leading edge material) to 300K at distances more than 50mm aft of the leading edge. These results clearly indicate that the wall thickness of the combustor and isolator need to be modified. As the temperatures in many parts of the engine exceed 600K, where the yield strength of Aluminum alloy is less than 18% of its room temperature value, a different material is required.

A thermal-structural analysis of the PDR scramjet intake (including leading edge) was repeated using low carbon steel. The temperature and von Mises stress at various positions along the intake (5mm thickness) as a function of time (altitude) are illustrated in Fig. **19a-b**. The temperature rise from stagnation point heating at the steel leading edge was also calculated for various wall thicknesses. The results suggest that if the intake wall thickness of the PDR model is increased, low carbon steel is a suitable material for the intake and the leading edge. Because a different material is not required for the leading edge and the rest of the intake, high thermal stresses at the interface, induced by the differences in thermal expansion coefficients, can be avoided.

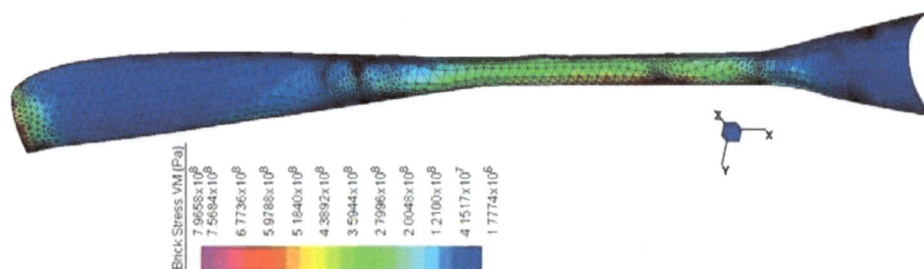

Figure 18: Temperature distribution in PDR model at 37km altitude, as taken from reference [8].

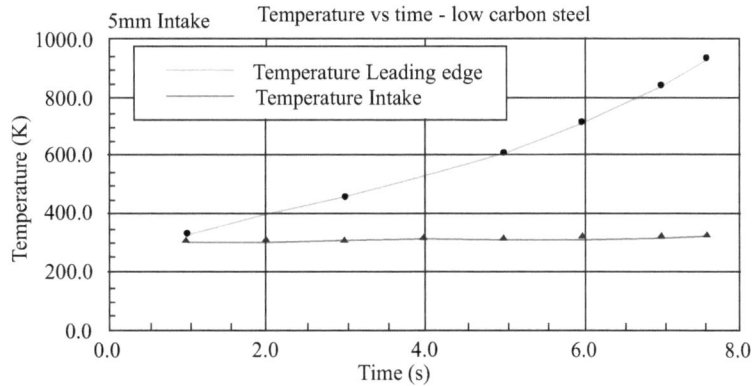

Figure 19a: Temperature vs. time (descent trajectory), as taken from reference [8].

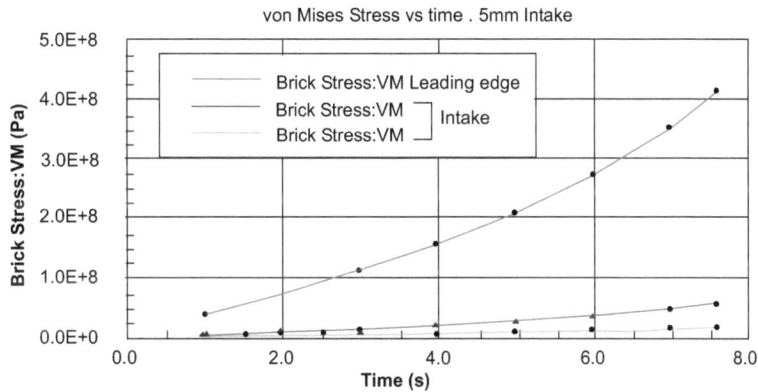

Figure 19b: Thermal induced stress vs. time, as taken from reference [8].

The effect of radiation to the atmosphere was also investigated. This was modelled by applying boundary conditions on the external surface of the structure. The radiation emissivities were varied from $\varepsilon = 0.1$ to 0.4. These values are typical for metals as received. Including the effect of radiation to the atmosphere during the descent trajectory made little difference to the results from the transient heat analysis.

Further details on the thermal-structural analysis of the Critical Design Review (CDR) models and the combustor can be found in reference [8].

5. FUTURE DIRECTIONS AND CHALLENGES

In recent years, significant advances in computational capabilities to numerically simulate and understand the physics of high-speed flows surrounding the vehicle to predict aerodynamic heating, the pressure and shear forces, and also the structural response to the aeroheating, aerodynamic and aeroacoustic loads have contributed to the proposal of more complex vehicle designs and missions in new international hypersonic initiatives (e.g. HyCause [50], HIFIRE [60]). However, the current state-of-the-art in the area of hypersonic aerothermoelasticity is not yet sufficiently adequate for the design of very complex types of hypersonic vehicles. The development of accurate and efficient CFD-based aerodynamic heating computations (e.g. incorporation of physics-based models for viscous and real-gas effects) will continue to be a critical research area.

It is expected that supersonic / hypersonic reuseable vehicles will have relatively long flight time in the atmosphere. This necessitates (in particular trans-atmospheric flights where severe, highly transient heat loads are generated) thermal management design considerations (e.g. efficient cooling concepts, thermal protection systems) and the use of exotic composite materials (e.g. intermetallic composites, functionally graded materials (FGM) thermal barrier coatings [61]) where modelling of the heat transfer and structural response can be very challenging. The aero-thermal-structural analysis of active cooling structures must consider heat transfer to and from both upper and lower surfaces of the vehicle over the flight regime as well as from the surface of the cooling system. Hence there is a need for more efficient and reasonably accurate coupled prediction methods suitable for design of these structures. There is also a requirement for nonlinear constitutive relations, which can take into account time and temperature dependence, for advanced composite materials such as intermetallics and FGMs. The use of advanced elevated temperature materials, actively cooled structures and heat exchangers also require the failure / damage tolerant models used in structural design and service life assessment to consider the combination of elevated temperature and environmentally induced degradation. For example, an actively cooled thermal structure could be exposed to the hydrogen / water engine exhaust and hydrogen coolant which may result in accelerated crack propagation, loss of ductility, accelerated creep behaviour and /or loss of damage tolerance.

Recent research [62] on adaptive materials showed that smart structures, which have the ability to change stiffness, shape, natural frequency, damping and other physical properties in response to changes in temperature, electric / magnetic fields, have great potential in many applications. Piezoelectric materials and shape memory alloys (SMAs) are some common examples of smart materials. It was noted in reference [63] that SMAs appear to have applications in high-performance aerospace vehicles in areas such as structural response control, structural shape control and damping enhancement. For example, smart structures could be used to increase the stability boundaries under combined aerodynamic and thermal loading conditions. A common / integrated platform is required in order to utilize the CFD data from the aerothermoelastic analysis of these smart structures in the control algorithms. Development of the ROM (Reduced Order Model) method [43] for performing aerothermoelastic solution, to yield data in a mathematical form that is suitable for use in the control and design environments, could advance the use of smart structures and adaptive control strategies for high-speed aerospace vehicles.

6. REFERENCES

[1] Anderson J.D. "Hypersonic and High Temperature Gas Dynamics", AIAA, Virginia, 2000.

[2] Bertin J.J. "Hypersonic Aerothermodynamics", AIAA education Series 1994.

[3] Ericsson LE. Unsteady Flows, Chapter 11 in "Tactical Missile Aerodynamics: General Topics", Edited by Hemsch MJ, Progress in Astronautics and Aeronautics 1992; Vol 141.

[4] Heisser WH, Pratt DT, Hypersonic Airbreathing Propulsion, AIAA Education Series, 1994.

[5] Ericsson LE, Reding JP. Asymmetric Flow Separation and Vortex Shedding on Bodies of Revolution, Chapter 10 in "Tactical Missile Aerodynamics: General Topics", Edited by Hemsch MJ, Progress in Astronautics and Aeronautics 1992; Vol 141.

[6] Hayes JR, Neumann RD, Chapter 3 in "Tactical Missile Aerodynamics: Prediction Methodology", Edited by Mendenhall EA, Progress in Astronautics and Aeronautics 1992; Vol 142.

[7] Ho SY, Paull A. Coupled Thermal, Structural and Vibrational Analysis of a Hpersonic Engine for Flight Test. Aeospace Science and Technology 2006; 10:420-426.

[8] Ho SY. Thermal-Structural Analysis of the DARPA HYCAUSE-3D Engine for Flight Test. 14th AIAA/AHI Space Planes and Hypersonic Systems and Technologies Conference 2006. AIAA 2006-8070.

[9] Ericsson LE, Almroth BO, Baile JA. Hypersonic Aerothermoelastic Characteristics of a Finned Missile. J.Spacecraft 1979; 16(3): 187-192

[10] Orlik-Rückermann KJ. Stability Derivatives of Sharp Wedges in Viscous Hypersonic Flow. AIAA Journal 1966; 4:1001-7.

[11] Ericsson LE, Viscous and Elastic Perturbations of Hypersonic Unsteady Airfoil Aerodynamics. AIAA Journal 1997; 15: 1481-90.

[12] Garrick LE. A Survey of Aerothermoelasticity. Aerospace Engineering 1963; 22:140-7.

[13] Hemsch MJ. Component Build-Up Method for Engineering Analysis of Missiles at Low-to-High Angles of Attack, Chapter 4 in "Tactical Missile Aerodynamics: Prediction Methodology", Edited by Mendenhall EA, Progress in Astronautics and Aeronautics 1992; Vol 142.

[14] Thuruthimattam BJ, Friedmann PP, McNamara JJ, Powell KG. Modeling Approaches to Hypersonic Aerothermoelasticity with Application to Reuseable Launch Vehicles. 44th AIAA/ASME/ASCE/AHS Structures, Structural Dynamics, and Materials Conference 2003; AIAA 2003-1967:1-19.

[15] McNamara JJ, Friedmann PP. Aeroelastic and Aerothermoelastic Analysis of Hypersonic Vehicles: Current Status and Future Trends. 48th AIAA/ASME/ASCE/AHS Structures, Structural Dynamics, and Materials Conference 2007; AIAA 2007-2013: 1-55.

[16] Wardlaw AB Jr, Baltakis FP, Priolo FJ, Solomon JM. Space Marching Euler Solvers. Chapter 9 in "Tactical Missile Aerodynamics: Prediction Methodology", Edited by Mendenhall EA, Progress in Astronautics and Aeronautics 1992; Vol 142.

[17] Lijewski LE, Simpson LB, Belk DM.Time-Asymptotic Euler, Navier-Stokes Methods for Subsonic/Transonic Flows. Chapter 10 in "Tactical Missile Aerodynamics: Prediction Methodology", Edited by Mendenhall EA, Progress in Astronautics and Aeronautics 1992; Vol 142.

[18] Ho SY. High Strain-Rate Consitutive Models. AIAA Journal of Propulsion and Power, 2002; 18(5):1106-1111.

[19] Rodgers JP. Aerothermoelastic Analysis of a NASP-Like Vertical Fin. Proceedings of the 33rd AIAA/ASME/ASCE/AHS Structures, structural Dynamics and Materials Conference 1992; AIAA-92-2400-CP.

[20] Rausch RD, Batina JT, Yang HT. Three-Dimensional Time-Marching Aeroelastic Analyses Using an Unstructured-Grid Euler Method. AIAA Journal 1993; 31(9):1626-1633.

[21] Edwards JW, Malone JB. A Current Status of Computational Methods for Transonic Unsteady Aerodynamics and Aeroelastic Applications. NASA Report 1991; NASA TM 104191.

[22] Bey KS, Thornton EA, Dechaumphai P, Ramakrishnan R. A New Finite Element Approach for Prediction of Aerothermal Loads – Progress in Inviscid Flow Computations. Presented at AIAA 7th Computational Fluids Dynamics Conference 1985;. AIAA Paper 85-1533CP.

[23] Thornton EA, Ramakrishnan R, Dechaumphai P. A Finite Element Approach for Solution of the 3D Euler Equations. Presented at the AIAA 24th Aerospace Sciences Meeting 1986. AIAA Paper 86-0106.

[24] Thornton EA, Dechaumphai P, Vemaganti G. A Finite Element Approach for Prediction of Aerothermal Loads. AIAA Paper 1986; AIAA 86-1050:1-11.

[25] Gupta KK. Development of a Finite Element Aeroelastic Capability. Journal of Aircraft 1996; 33(5): 995-1002.

[26] Hayes JR, Neumann RD. Introduction to the Aerodynamic Heating Analysis of Supersonic Missiles, Chapter 3 in "Tactical Missile Aerodynamics: Prediction Methodology", Edited by Mendenhall EA, Progress in Astronautics and Aeronautics 1992; Vol 142.

[27] Quinn RD, Gong L. A Method for Calculating Transient Surface Temperatures and Surface Heating Rates for High-Speed Aircraft. NASA Report 2000; NASA/TP-2000-209034.

[28] Ho SY, Carè G. Modified Fracture Mechanics Approach in Structural Analysis of Solid Rocket Motors. Journal of Propulsion and Power 1998; 14(4): 409-415.

[29] Codrington J, Nguyen P, Ho SY, Kotousov A. Induction Heating Apparatus for High Temperature Testing of Thermo-Mechanical Promerties. Applied Thermal Engineering 2009; in press.

[30] MIL-STD Composite Handbook MIL-HDBK-5H, 1998.

[31] Shidelar JL, Webb GL, Pittman CM. Verification Tests of Durable Thermal Protection Systems Concepts. Journal of Spacecraft and Rockets 1985; 22(6), 598-604.

[32] Thornton EA, Dechaumphai P. Coupled Flow, Thermal, and Structural Analysis of Aerodynamically Heated Panels. Journal of Aircraft 1988; 25:1052-1059.

[33] Dryden HL, Duberg JE. Aeroelastic Effects of Aerodynamic Heating. Proceedings of the Fifth AGARD General Assembly, 1955.

[34] Tran H, Farhat C. An Integrated Platform for the Simulation of Fluid-Structure-Thermal Integration Problems. AIAA/ASME/ASCE/AHS/ASC Structures, Structural Dynamics and Materials Conference 2002, AIAA 2002-1307.

[35] Silva WA, Bartels RE. Development of Reduced-Order Models for Aeroelastic and Flutter Prediction Using the CF3Dv6.0 Code. American Institute of Aeronautics and Astronautics Paper 2002; AIAA 2002-1596.

[36] Heeg J, Gilbert MG, Pototzky AS. Active Control of Aerothermoelastic Effects for a Conceptual Hypersonic Aircraft. Journal of Aircraft 1993; 30(4): 453-458.

[37] Morgan HG, Runyan HL and Huckel V. Theoretical Considerations of Flutter at High Mach Numbers. Journal of the Aeronautical Sciences 1958; 25(6): 371-381.

[38] Lohner R, Yang C, Cebral J, Baum J. Fluid-Structure-Thrmal Interaction Using a Loose Coupling Algorithm and Adaptive Unstructured Grids. Proceedings of the 29th AIAA Fluid Dynamics Conference 1998; AIAA 98-2419.

[39] Cary AM, Bertram MH. Engineering Prediction of Turbulent Skin Friction and Heat Transfer in High-Speed Flow. NASA Report 1974; NASA TN D-7507.

[40] Paull A. Fundamental Research Challenges and Opportunities Related to HIFIRE and beyond. Proceedings of the Australian Hypersonics Capabilities and Future Directions Workshop 2008.

[41] Atluri SN. Computational Non-Linear Mechanics in Aerospace Engineering. Progress in Astronautics and Aeronautics 1992; Volume 146.

[42] Silva WA. Recent Enhancements to the Development of CFD-based Aeroelastic Reduced-Order Models. Proceedings of the 48th AIAA/ASME/ASCE/AHS/ASC Structures, Structural Dynamics, and Materials Conference 2007; AIAA 2007-2051.

[43] Silva WA. Simultaneous Excitation of Multiple-Input Multiple-Output CFD-Based Unsteady Aerodynamic Systems. American Institute of Aeronautics and Astronautics 2007; AIAA Paper 2007-XXX.

[44]   Paull A, Alesi H, Anderson S. The HyShot Flight Program. Proceedings of the 11th AIAA/AAAF Space Planes and Hypersonic Systems and Technologies Conference, 2002.

[45]   Abdel-Jaward M, Paull A. Design of a 2D Intake for a Scramjet. Proceedings of the 9th International Workshop on Shock Tube Technology, 2004.

[46]   Heiser WH, Pratt DT. Hypersonic Airbreathing Propulsion, AIAA Education Series, 1994.

[47]   Eckert ERG. Engineering Relations for Friction and Heat Transfer to Surfaces in High Velocity Flow, Reader's Forum 1995; 565.

[48]   Ho SY. Prediction and Control of Aerothermoelastic Effects in Hyersonics Vehicles. Proceedings of the Australian Hypersonics Capabilities and Future Directions Workshop 2008.

[49]   MIL-STD Composites Handbook MIL-HDBK-5H 1998.

[50]   Walker S, Rodgers F. Falcon Hypersonics Technology Overview. Proceedings of the 13th International Space Planes and Hypersonic Systems and Technologies Conference 2005; AIAA 2005-3253.

[51]   Holden MS, Wadhams TP, Smolinski GJ, Macleans MG, Harvey J, Walker BJ. Experimental and Numerical Studies on Hypersonic Vehicle Performance in LENS Shock and Expansion Tunnels. Proceedings of the 44th AIAA Aerospace Sciences Meeting and Exhibit, 2006. AIAA 2006-125.

[52]   Blake WB. Missile Datcom Users Manual-1998 Fortran 90 Revision 1997; AFRL-VA-WP-TR-1998-3009.

[53]   Ericsson LE. Viscous and Elastic Perturbations of Hypersonic Unsteady Airfoil Aerodynamics. AIAA Journal 1977; 15(10): 1481-1490.

[54]   Dechaumphai P, Morgan K. Transient Thermal-Structural Analysis Using Adaptive Unstructured Remeshing and Mesh Movement. Chapter in "Thermal Structures and Materials for High-Speed Flight", Edited by Thornton EA, Progress in Astronautics and Aeronautics 1992; Vol 140.

[55]   Bodner SR. Review of a Unified Elastic-Viscoplastic Theory. Chapter in "Unified Constitutive Equations for Creep and Plasticity", Edited by Miller AK 1987.

[56]   Bodner SR, Partom Y. Constitutive Equations for Elastic-Viscoplastic Strain-Hardening Materials. Journal of Applied Mechanics 1975; 42: 385-389.

[57]   Pandey AK, Dechaumphai P, Thornton EA. Finite Element Thermoviscoplastic Analysis of Aerospace Structures. Chapter in "Thermal Structures and Materials for High-Speed Flight", Edited by Thornton EA, Progress in Astronautics and Aeronautics 1992; Vol 140.

[58]   Van Stone RH, Kim KS. Methods for Predicting Crack Growth in Advanced Structures. Chapter in "Thermal Structures and Materials for High-Speed Flight", Edited by Thornton EA, Progress in Astronautics and Aeronautics 1992; Vol 140.

[59]   Chamis CC, Shiao M. Probabilistic Assessment of Thermal Structures. Chapter in "Thermal Structures and Materials for High-Speed Flight", Edited by Thornton EA, Progress in Astronautics and Aeronautics 1992; Vol 140.

[60]   Dolvin D. HIFIRE Overview. Proceedings of the Australian Hypersonics Capabilities and Future Directions Workshop 2008.

[61]   Tsukamoto H, Kotousov A, Ho SY, Codrington J. Analysis and Design of Functionally Graded Thermal Coating. Proceedings of the Structural Integity and Failure Conference 2006; 25-32.

[62]   Oh IK, Han JH, Lee DM. Postbuckling and Vibration Characteristics of Piezolaminated Composite Plate Subjected to Thermo-Piezoelectric Loads. J Sound Vibration 2000; 233: 19-40.

[63]   Lee I, Roh JH, Oh IK. Aerothermoelastic Phenomena of Aerospace and Composite Structures. Journal of Thermal Stresses 2003; 26:525-546.

[64]   Lighthill M. Oscillating Airfoils at High Mach Number. Journal of the Aeronautical Sciences1953; 20(6): 402-406.

[65]   Asheley H, Zartarium G. Piston Theory – A New Aerodynamic Tool for the Aeroelastician. Journal of the Aeronautical Sciences 1956; 23(12): 1109-1118.

[66]   Van Dyke M.D. The Supersonic Blunt Slender Body Problem – Review and Extension. Journal of the Aero Space Sciences 1958; 25:485-495.

[67]   Oppenheimer M.W., Doman D.B. A Hypersonic Vehicle Model Developeed with Piston Theory. Proceedings of the 2006 AIAA Atmospheric Flight Mechanics Conference 2006; AIAA-2006-6637.

Structural Failure Analysis and Prediction Methods for Aerospace Vehicles and Structures, 2010, 85-95

CHAPTER 4

Fatigue Crack Growth Analysis for Notched Specimens under Flight Spectrum Loading

W. Zhuang and L. Molent

Air Vehicles Division, Defence Science and Technology Organisation, Melbourne, VIC 3207, Australia

Abstract: In aerospace vehicle design, structures always contain a range of geometric variations including various notches, holes and cutouts to achieve certain mechanical functions as well as weight savings. Those notches and holes can cause local stress concentrations and/or create hot spots for fatigue crack initiation and propagation. This chapter presents an analytical study of fatigue crack growth in aluminium alloy 7050-T7451 notched specimens under a fighter aircraft wing root bending moment spectrum. The crack growth data were measured by quantitative fractography for three groups of specimens with different stress concentration geometrical features. Under spectrum loading and for each spectrum peak stress level, a minimum of five specimens were tested. Based on the analysis of the measured spectrum crack growth data using linear elastic fracture mechanics, it was found that the concept of geometry factors formulated in the stress intensity factor could not collapse the crack growth rate data derived from each stress concentration feature, particularly near the small crack growth region. In order to investigate the possible reasons for this, three-dimensional elastic-plastic finite element analysis was used to determine notch plastic zone sizes for each stress concentration geometry. As a consequence, an alternative crack growth driving force which considered both notch elastic-plastic stress field and gross net-section stress field was used to interpret the fatigue crack growth data under spectrum loading. It was found that the predictions of crack growth under spectrum loading for different stress concentration factors at different peak load levels agreed reasonably well with the experimental results.

INTRODUCTION

In aerospace vehicle design, structures always contain a range of geometric variations including various notches, holes and cutouts to achieve certain mechanical functions and to reduce aircraft weight. Those notches and holes can cause local stress concentrations and may promote fatigue crack initiation and propagation [1-3]. During the operational life of aircraft, its fatigue critical structures and components are subjected to variable amplitude flight spectrum loading [4, 5]. Therefore, the objective of this chapter is to investigate and analyze fatigue crack growth (FCG) behavior in aluminium alloy specimens with different stress concentration features, subjected to variable amplitude loading.

For predicting the fatigue life of notched structures, engineers and scientists often find themselves in a situation where FCG data were obtained experimentally from specimens or structures with different stress concentration factors (K_t) or even only from smooth specimens. How to predict the fatigue life of the notched structures, based on these existing FCG database, is still a challenging topic. Therefore, an extensive research effort [e.g. 6-14] has been devoted to investigating the effect of geometric discontinuities, namely notches on FCG behavior due to their importance in aircraft structural damage tolerance analysis. It is noteworthy that almost all these studies were carried out based on constant-amplitude (CA) fatigue tests at different stress ratios (R), except the study presented in ref [14]. For instance, from some experimental results in [13], Fig. **1** shows the FCG behavior for different U-shaped notch depths (D) in a low carbon steel plate for the same notch root radius r=1.0 mm under CA loading at R=0. It indicated that FCG rates are higher for the deeper notch. Fig. **2** shows the notch root radii effect on FCG behavior of the same U-shaped notch depth D=2.0 mm specimens under CA loading at R=0. It was found that FCG rate is higher for blunter notch r=1.0 mm than that for r=0.5 mm. Most importantly, the results demonstrated a similar tendency that the crack growth rate decreases initially to a minimum value before it accelerates. The minimum value in the crack growth rate or the "V" shaped curve of crack growth near a notch is often characterized as the anomalous behavior of short FCG attributed to a lack of crack closure [15].

Shin *et al.* [11] claimed a similar crack growth behavior in BS1470 S1C aluminium alloy single edge notched specimens (420 mm × 140 mm × 3 mm) with various notch root radii as shown in Fig. **3**. The specimens with a constant semi-elliptical notch depth of 35 mm were tested under CA loading with $\Delta\sigma_{app} = 15 MPa$ and $R = 0.05$. For the notch root radii r=1.4 mm, however, fatigue crack growth did not follow the "V" shaped curve near the notch. It implies some complicated mechanisms influencing the growth behavior of cracks from notches. It seems that the behavior for the notch free case in Fig. **3** is particularly related to short crack growth.

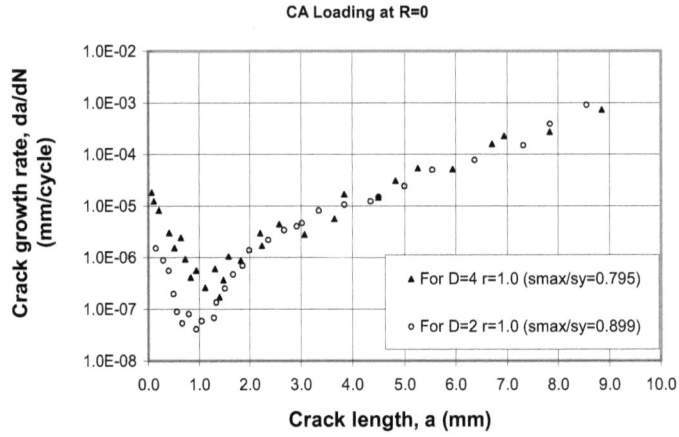

Figure 1: FCG in BS4360 50B steel with different U-notch depth [13].

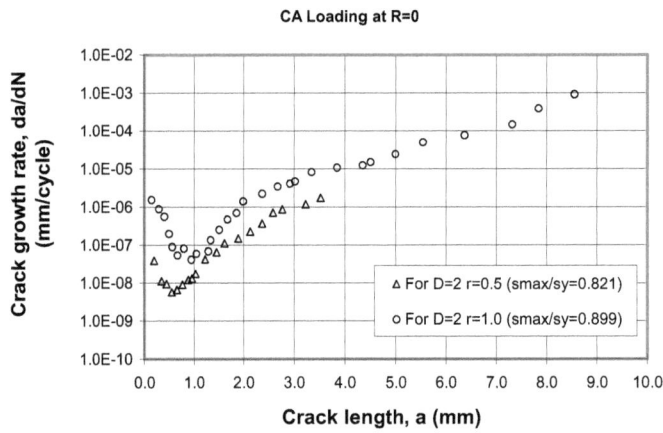

Figure 2: FCG in BS4360 50B steel with different U-shaped notch root radii.

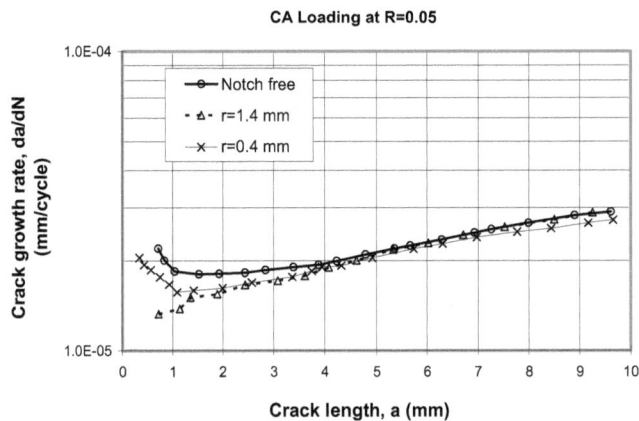

Figure 3: FCG in BS1470 S1C aluminium alloy specimens with different notch root radii.

Using the linear elastic fracture mechanics (LEFM) approach, El Haddad *et al.* [9] analyzed the crack growth data in a G40.11 steel for different geometric discontinuities subjected to a far field applied stress range $\Delta\sigma_{app} = 269\,MPa$ and $R = -1$. They discovered that crack growth rates for circular and elliptical hole specimens are initially higher than that for the central cracked plate as shown in Fig. **4**. The crack length in Fig. **4** includes the notch depth. From the findings, they indicated that the growth rate difference gradually reduces and eventually disappears when fatigue cracks propagate out of the notch inelastic stress-strain fields. Most importantly, they concluded that LEFM fails to characterize FCG driving force by a unique mathematical relationship or equation. In some cases, short cracks from notches also grow at the levels below the crack growth threshold values determined by conventional LEFM tests [16].

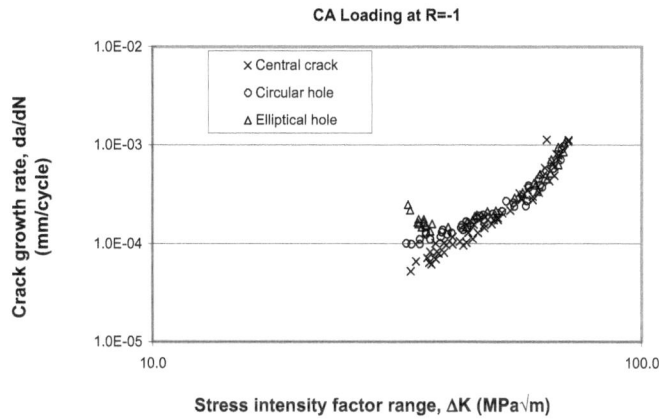

Figure 4: Fatigue crack growth in G40.11 steel specimens for different geometric features.

For high strength aluminium alloys such as 7075-T6 [12], small FCG data from a single edge notch tensile specimen were measured by the plastic-replica method. Fig. **5** shows the small crack growth rates under constant amplitude loading at *R*=−1. It was found that the small crack effects are much more pronounced for this high strength aluminium alloy. It means that small cracks can grow at *ΔK* levels well below the large crack threshold ΔK_{th} and also propagate faster than large cracks at the same *ΔK* level. However, the paper did not investigate the influence of notch plasticity on small FCG emanating from the notches.

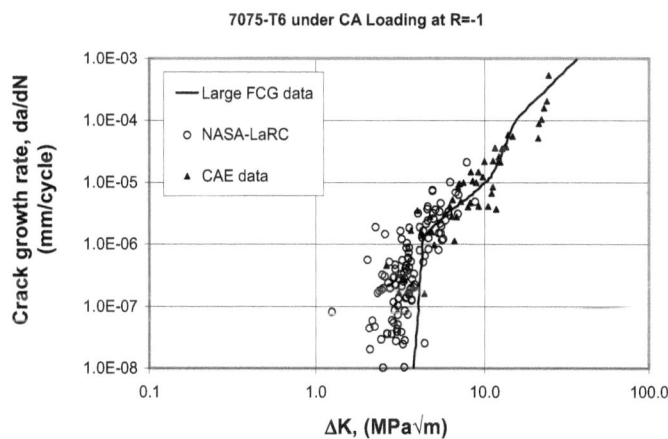

Figure 5: Small and large crack growth rates under constant amplitude loading at R=−1 for 7075-T6.

In general, depending on notch geometry, material and loading conditions, it is believed that there are mainly two mechanisms dominating the growth behavior of fatigue cracks from notches: (a) notch plasticity and (b) material microstructures [10]. Crack retardation at microstructural interfaces is one of mechanisms used to explain the initially decreasing growth rates for short cracks. Based on fracture mechanics, however, several approaches have been developed to consider the notch plasticity effect. For example, by modifying the stress intensity factor as a function of not only applied stress and crack length but also notch stress concentration factor and notch tip radius, Kujawski [17] successfully correlated FCG data from notches experimentally obtained by Smith *et al.* [7] using this approach. However, more recently, for CA loading, Ding *et al.* [18] found that the effect of the notch plasticity on FCG happened only when $R>0$. It was argued that when the R-ratio was negative, the contact of the crack surfaces during a part of an unloading cycle reduced the cyclic plasticity of the material near the crack tip. The combined effect of notch plasticity and possible contact of the crack's surface were claimed responsible for the observed crack growth phenomenon near a notch.

In addition to the modified ΔK approach, there are several other elastic-plastic fracture mechanics (EPFM) approaches such as modified J-integral [19], crack tip opening displacement (CTOD) [20], crack tip opening angle (CTOA) [21], and cyclic notch plasticity [22] used in an attempt to correlate the experimental FCG data. However, for the application of these approaches to interpret the experimentally-obtained short FCG data from notched specimens, many studies [11, 13, 23] indicated difficulties such as limited data and computational burden experienced by using these approaches. It was also found that it is difficult to determine the J-integral, CTOD and CTOA experimentally and numerically when the crack length is small. For instance, CTOA is mainly used to correlate FCG data for large fatigue crack growth in the pipeline industry.

FATIGUE CRACK GROWTH UNDER SPECTRUM LOADING FOR DIFFERENT STRESS CONCENTRATIONS

Fatigue crack growth tests of notched specimens under spectrum loading conditions were conducted at DSTO. The following sections give brief details of the test specimens, material properties, spectrum loading and test procedure. Full details can be found in references [24-26].

Geometries of Notched Specimens

Three groups of specimens with different net-section stress concentration factors (K_{tn}) were manufactured as shown in Figs. **6-8**. The net-section stress concentration factors were calculated by the finite element method (FEM) that will be described in the subsequent section.

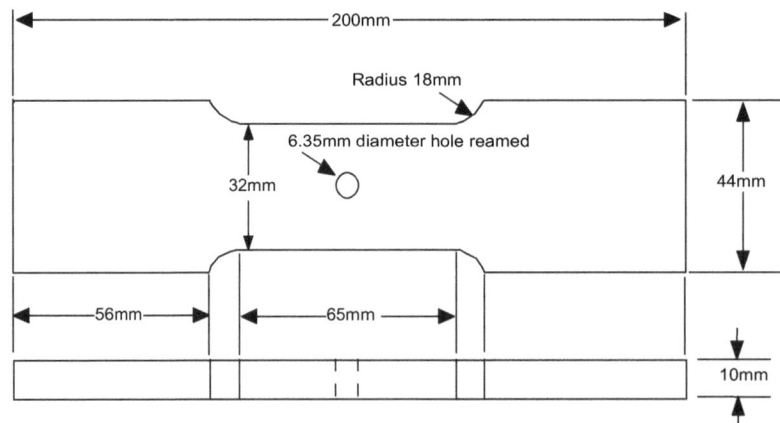

Figure 6: The high K_t specimen (K_{tn}=2.68).

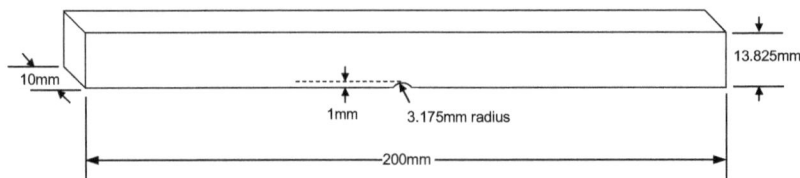

Figure 7: The mid K_t specimen (K_{tn}=2.13).

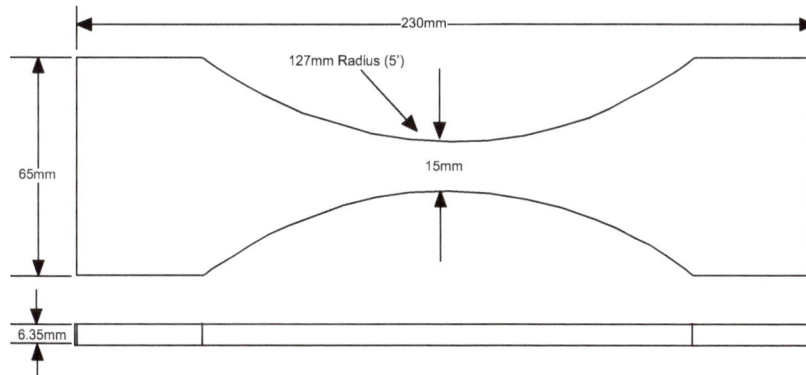

Figure 8: The low K_t specimen (K_{tn}=1.06).

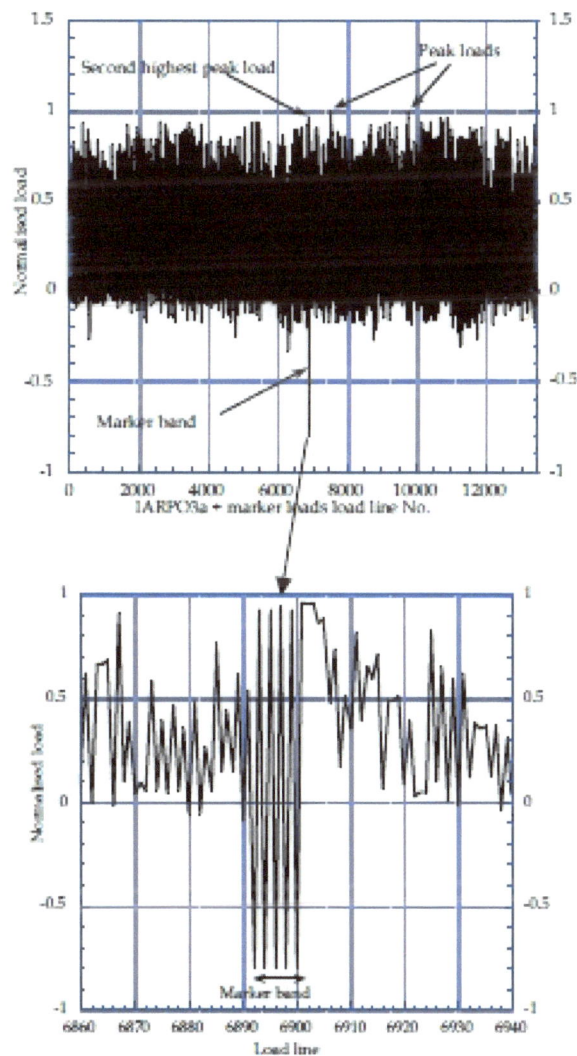

Figure 9: A modified version of a fighter aircraft wing root bending moment spectrum with marker-loads.

Material Properties of 7050-T7451

All specimens were made from thick plate 7050-T7451 aluminium alloy. The general mechanical properties of 7050-T7451 are listed in Table **1** [27].

Table 1: Mechanical properties of 7050-T7451.

Property	Value
Young's modulus	71000 [MPa]
Tensile Strength	482.6 [MPa]
Yield Strength	413.7 [MPa]
Poisson's ratio	0.33
Fracture toughness	35.71 [MPa$\sqrt{m}$]

Spectrum FCG Testing

All specimens were subjected a modified version of a fighter aircraft wing root bending moment spectrum. Marker-loads indicated in Fig. **9** were inserted into the spectrum just before one of the highest loads and between the four preceding tensile loads to facilitate the Quantitative Fractography (QF) method for the measurement of the crack growth [24].

Notch Plastic Zone Sizes of the Specimens

Three-dimensional (3D) elastic-plastic finite element analysis was used to determine notch plastic zone sizes for each stress concentration geometry such as that for the high K_t specimen in Fig. **10**.

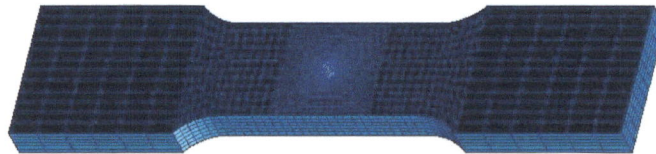

Figure 10: The 3D finite element mesh for high K_t.

Notch plastic zone sizes (NPZS) along the crack plane for the three types of specimens were calculated as shown in Fig. **11** for high K_t at the peak stress level of 250 MPa.

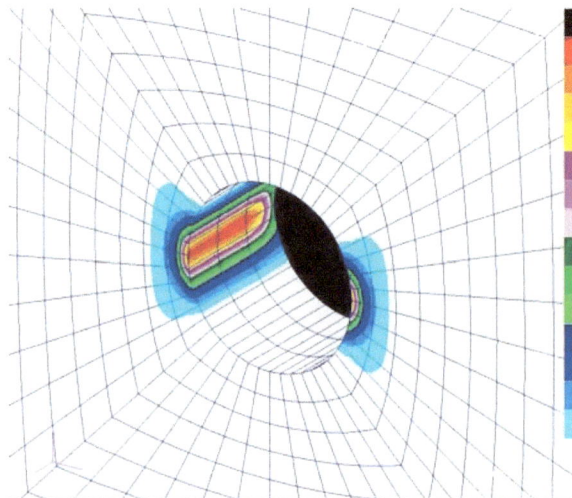

Figure 11: Plastic strain contour for high K_t.

Fig. **12** shows the curves of NPZS against applied maximum net-section stresses for low and high K_t specimens. It is noteworthy that the low K_t specimens at the peak stress levels of 420 and 450 MPa experienced gross net-section yielding.

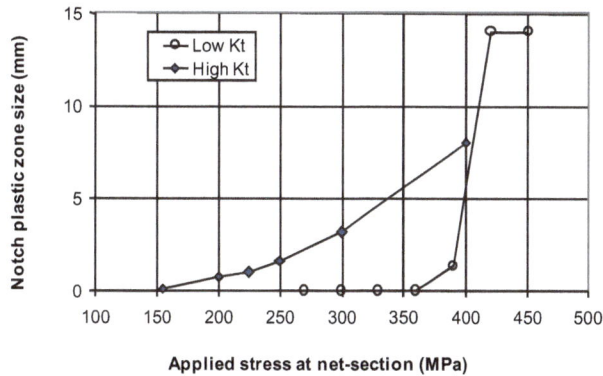

Figure 12: Notch plastic zone sizes vs applied stresses for low and high K_t.

ANALYSIS OF SPECTRUM FCG FOR DIFFERENT STRESS CONCENTRATION FACTORS

For all three types of specimens under the spectrum loading, FCG data were measured by QF. Huynh *et al.* [26] analyzed the FCG data using the LEFM approach and found that the conventional geometry factors failed to collapse the crack growth rate data derived from each stress concentration features as shown in Fig. **13**.

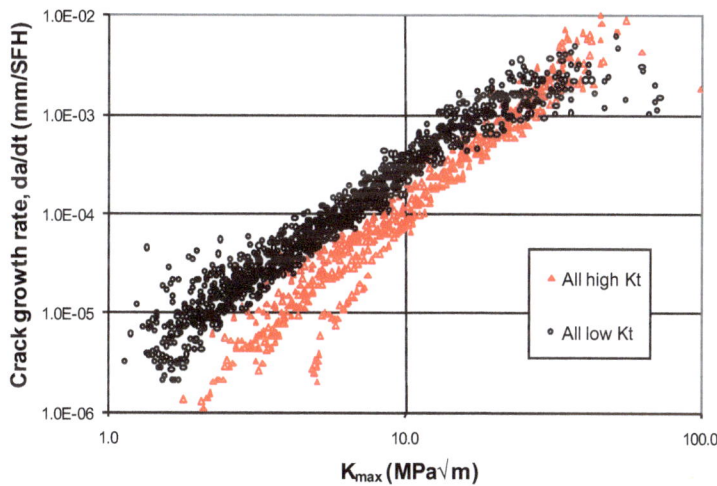

Figure 13: Paris-type FCG curves (Fig.14 in [27]).

Fig. **14** is the curves (open symbols) of crack growth rate vs crack length from the notch root of the high K_t specimens at the peak net-section stress of 155 MPa. As can be seen, similar to FCG from notches under the CA loading such as that in Fig. **3**, there are also two "V" shaped curves when cracks are small near the notch root under the spectrum loading. But, the other three curves seem to be uninfluenced by the notch root plastic zone. More importantly, it demonstrates for all five high specimens at the higher peak stress level of 250 MPa, that crack growth rates increase proportionally with crack size.

Considering that current engineering practice of damage tolerance analysis for aircraft structures is still mainly based on the LEFM approach [28], in this study, it is proposed that the crack driving force is defined by two local stress fields at net-section to account for the notch plasticity effect. The first stress field is local notch stress state dominated by NPZS while the other is gross net-section stresses. It is postulated that the crack growth rate da/dt is written as a function of the local notch stress intensity factor K_{n_peak}.

$$da/dt = C(K_{n\_peak})^m \tag{1}$$

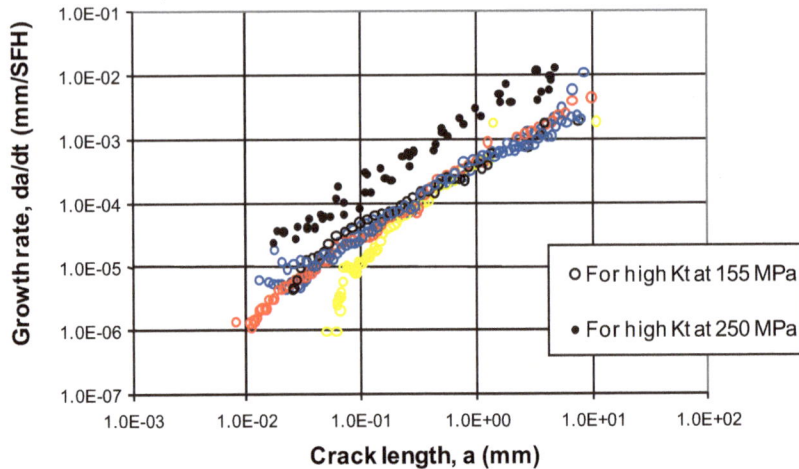

Figure 14: FCG rate vs crack length for high K_t at the peak stress of 155 and 250 MPa.

where the constants C and m are experimentally derived. It is believed that they should be the functions of local stress distributions at notches for given material and loading condition. For spectrum loading, the peak notch stress intensity factor is defined according to NPZS as follows

$$K_{n\_peak} = \sigma_o \sqrt{\pi a} \text{ for } a \leq NPZS \tag{2a}$$

$$K_{n\_peak} = \beta \sigma_{n\_peak} \sqrt{\pi a} \text{ for } a > NPZS \tag{2b}$$

where σ_o is the material flow stress for the elastic-perfectly plastic condition, σ_{n_peak} is the peak net-section stress in the spectrum block loading and β is the geometry factor.

Fig. 15 is the FCG rate curves plotted against the crack growth driving force, K_{n_peak}. It appears that the proposed approach is able to significantly collapse the crack growth rate data for all high and low K_t specimens.

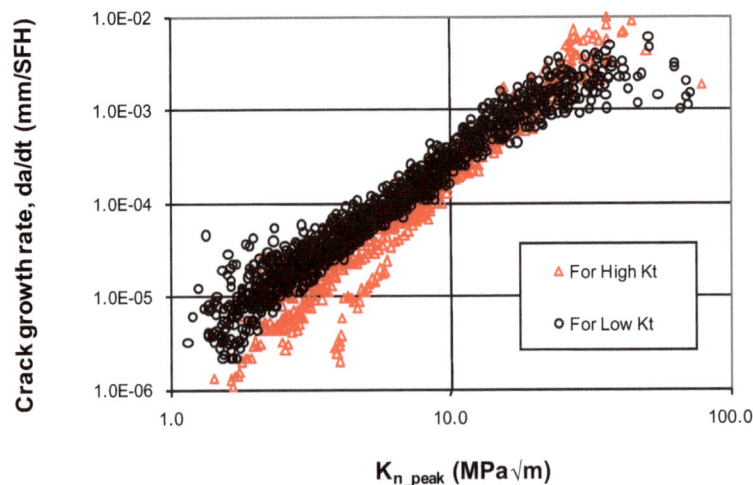

Figure 15: FCG rate vs K_{n_peak} for all high and low K_t specimens.

Fig. 16 presents the effect of peak net-section stress on the FCG exponent m for both low and high K_t specimens by the linear regression of Eq. (1). It was found that the exponent is approximately equal to two under spectrum loading, not only for small-scale yielding but also for gross-section yielding conditions. The findings are consistent with the results presented in [26]. It simplifies the prediction of the unknown parameter C, which was found to be a function of stress concentration factors and net-section peak stress level as shown in Fig. 17, for given material and loading condition.

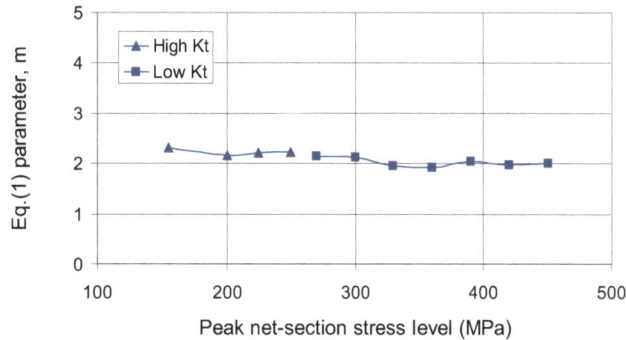

Figure 16: The effect of K_t and σ_{n_peak} on m.

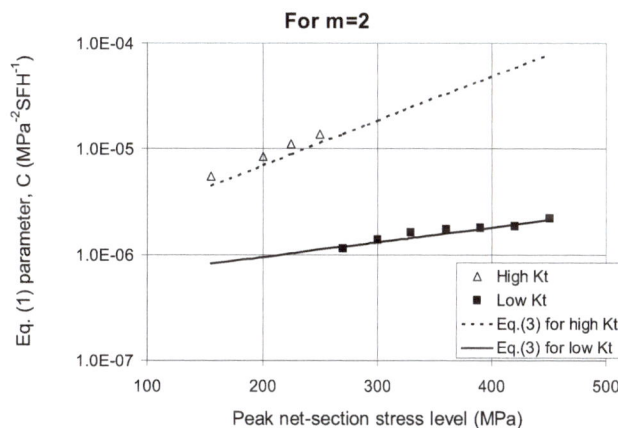

Figure 17: The effect of K_t and σ_{n_peak} on C.

For the same spectrum, from Fig. 17, an empirical equation can be derived to predict the unknown parameter C for different stress concentration factors at different net-section stress as follows:

$$C = e^{A_0 + A_1 K_t + A_2 peak} \tag{3}$$

where A_0, A_1 and A_2 are constants determined by multiple regression. From the low and high K_t FCG datasets, the constants are -1.71E+01, 1.6E+00 and 6.40E-03 respectively. To validate Eqs. (1)-(3), the mid K_t FCG data were used. Fig. **18** demonstrated that the predictions of FCG for the mid K_t specimens at the net peak levels of 180 and 200 MPa are consistent with the experiments. It also indicated that both tested and predicted FCG rates for the mid K_t specimens are weakly stress dependent.

CONCLUSIONS

This chapter presented an analysis of fatigue crack growth from notched specimens subjected to flight spectrum loading. An alternative crack growth driving force considering both the notch elastic-plastic stress field and gross net-section stress field was proposed to interpret spectrum fatigue crack growth data with three different stress concentration factors.

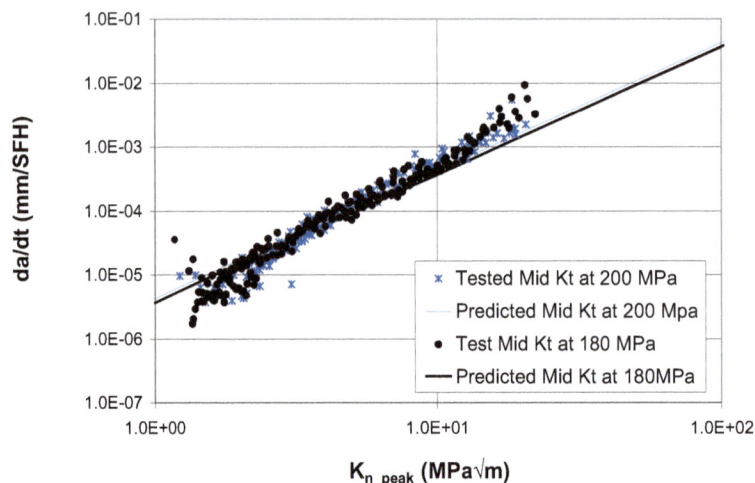

Figure 18: Validation of FCG prediction for Mid K_t.

The notch elastic-plastic stresses under cyclic loading were determined by three-dimensional elastic-plastic finite element method. Notch plastic zone sizes for each stress concentration geometry were subsequently calculated and used for the FCG analysis.

The parameters of the FCG prediction model were derived based on low and high K_t FCG data. Mid K_t FCG data was used to validate the model. The results demonstrated that the predictions of FCG for different stress concentration factors (such as for the mid K_t specimens) at different load levels agreed reasonably well with the experimental results. It is believed that this work could lead to a better prediction of FCG life for airframes containing different geometric discontinuities.

REFERENCES

[1]   Zhuang W, Barter S, McDonald M, Molent L. Fatigue crack growth life assessment of the F/A-18 aft fuselage frames (Y645 and Y657). *DSTO-TR-1826,* Defence Science and Technology Organisation, Melbourne, Australia; Feb 2006.

[2]   Zhuang W, Barter S, Molent L. Flight-by-flight fatigue crack growth life assessment. *International Journal of Fatigue* 2007; Vol. 29: 1647–1657.

[3]   Zhuang, W., Molent, L. Block-by-block approaches for spectrum fatigue crack growth prediction, *Engineering Fracture Mechanics* 2008; Vol.75 (17): 4933-4947.

[4]   Schijve J. Fatigue damage in aircraft structures, not wanted, but tolerated? *International Journal of Fatigue* 2009; Vol. 31: 998-1011.

[5]   Molent L. The history of structural fatigue testing at Fishermans Bend Australia. *DSTO-TR-1773,* Defence Science and Technology Organisation, Melbourne, Australia; Oct 2005.

[6]   Frost N, Dugdale D. Fatigue tests on notched mild steel plates with measurements of fatigue cracks, *Journal of the Mechanics and Physics of Solids* 1957; Vol. 5: 182-192.

[7]   Smith R, Miller K. Fatigue cracks at notches. *International Journal of Mechanical Science* 1977; Vol. 19: 11–22.

[8]   Dowling N. Notched member fatigue life predictions combined crack initiation and propagation, *Fatigue of Engineering Materials and Structures* 1979; Vol. 2: 129–138.

[9]   El Haddad M, Smith K, Topper T. A strain based intensity factor solution for short fatigue cracks initiating from notches, *Fracture Mechanics, ASTM STP 677, C. Smith, Ed.,* American Society for Testing and Materials, 1979; 274-289.

[10]   Leis B. Displacement controlled fatigue crack growth in inelastic notch fields: Implications for short cracks, *Engineering Fracture Mechanics* 1985; Vol. 22 (2): 279-293.

[11]   Shin C, Smith R. Fatigue crack growth at stress concentrations – The role of notch plasticity and crack closure, *Engineering Fracture Mechanics* 1988; Vol. 29 (3): 301-315.

[12]   Wu X. R., Newman J. C., Zhao W., Swain M. H., Ding C. F. and Phillips E. P., Small crack growth and fatigue life predictions for high-strength aluminium alloys: part I -experimental and fracture mechanics analysis, *Fatigue & Fracture of Engineering Materials & Structures* 1998; Vol.21: 1289–1306.

[13] Hammouda M, Osman H, Sallam H. Mode I notch fatigue crack growth behaviour under constant amplitude loading and due to the application of a single tensile overload, *International Journal of Fatigue* 2004; Vol. 26: 183–192.

[14] Zhuang W. and Molent L. Analysis of spectrum fatigue crack growth in AA7050 specimens with three stress concentrations, *Proceedings of International Conference on Crack Paths,* Vicenza, Italy, September 2009.

[15] Pluvinage G. Fracture and fatigue emanating from stress concentrators. Kluwer Academic Publishers, 2003.

[16] Pan J, De Los Rios E, Miller K. Short fatigue crack growth in plane and notched specimens of an 8090 Al-Li alloy, *Fatigue and Fracture of Engineering Materials and Structures* 1993; Vol. 16 (12): 1365-1379.

[17] Kujawski D. Estimations of stress intensity factors for small cracks at notches, *Fatigue and Fracture of Engineering Materials and Structures* 1990; Vol. 14 (10): 953-965.

[18] Ding F, Feng M, Jiang Y. Modeling of fatigue crack growth from a notch, *International Journal of Plasticity* 2007; Vol. 23: 1167-1188.

[19] Chen Y, Lu T. On the path dependence of the J-integral in notch problems, *International Journal of Solids and Structures* 2004; Vol. 41: 607-618.

[20] Hammouda M, Ahmad S, Sherbini A, Sallam H. Deformation behaviour at the tip of a physically short fatigue crack due to a single overload, *Fatigue and Fracture of Engineering Materials and Structures* 1999; Vol. 22 (2): 145-151.

[21] Darcis P, Drexler E, Fields R, McColskey J., McCowan C., Reuven R. and Siewert T. Crack tip opening angle: Applications and developments in the pipeline industry, *Proceedings of the 12th International Conference on Fracture*, Ottawa Canada, July 2009.

[22] Li W. Short fatigue crack propagation and effect of notch plastic field, *Nuclear Engineering and Design* 2003; Vol. 84: 193-200.

[23] Navarro A, De los Rios E. Fatigue crack growth modelling by successive blocking of dislocations, *Proc. R. Soc. Lond. A* 1992; Vol. 437: 375-390.

[24] Barter S, Huynh J. Fatigue crack growth in 7050-T7451 aluminium alloy open hole coupons, *DSTO-TN-0677*, Defence Science and Technology Organisation, Melbourne, Australia, Jan 2006.

[25] Huynh J, Barter S. Fatigue crack growth in 7050-T7451 aluminium alloy notched coupons, *DSTO-TR-1966*, Defence Science and Technology Organisation, Melbourne, Australia, March 2007.

[26] Huynh J, Molent L, Barter S. Experimentally derived crack growth models for different stress concentration factors, *International Journal of Fatigue* 2008; Vol. 30: 1766-1786.

[27] The Federal Aviation Administration, Metallic Materials Properties Development and Standardization (MMPDS-3), USA. October 2006.

[28] Ball D, Norwood D, TerMaath S. Joint strike fighter airframe durability and damage tolerance certification. In: *Proceedings of the 47th AIAA/ASME/ASCE/AHS/ASC struct. dyn. mater. conf.* Rhode Island, USA; May 2006.

96   *Structural Failure Analysis and Prediction Methods for Aerospace Vehicles and Structures,* 2010, 96-132

CHAPTER 5

Application of Refined Plate Theory to Fracture and Fatigue

A. Kotousov and J. Codrington

School of Mechanical Engineering, the University of Adelaide, Australia

Abstract: The work presented here is a compendium of theoretical results obtained by the authors between 2005 and 2009. Among these results are comprehensive analysis of the three-dimensional elastic stress and displacement fields near a tip of a through-the-thickness crack, generalization of the classical strip-yield model for plates having a finite thickness, and development of an analytical approach for calculating the plasticity-induced crack closure and crack growth rates at constant and variable amplitude loading. As an application of the developed approach, new predictive models of various non-linear fatigue crack growth phenomena in plates of finite thickness were developed. These include computational models of crack growth under small-scale yielding conditions and constant amplitude loading, growth of a fatigue crack emanating from a sharp notch, and crack growth retardation phenomenon following an overload cycle. All theoretical predictions were extensively compared with previous numerical and experimental studies demonstrating a great potential of the refined plate theory in the analysis of fracture and fatigue problems.

INTRODUCTION

Most analytical and numerical studies in fracture mechanics over the past fifty years were based on two-dimensional solutions of the classical plane theory of elasticity, though three-dimensional effects were often acknowledged. Relative simplicity is the main reason for the popularity of these solutions as three-dimensional formulations are not very amendable to analytical techniques.

The plane solutions accommodate two main assumptions regarding the state of stress in a plate subjected to in-plane loading: plane stress (zero transverse stresses) and plane strain (zero transverse strain components). However, so far there is no generally accepted criterion for identifying what thickness would qualify as plane-stress or plane-strain and, in general, what effect on the stress distribution the plate thickness has. More importantly, many numerical and experimental studies have demonstrated that the actual displacement and stress distributions in the vicinity of a tip of a crack do not follow either plane stress or plane strain assumptions. Therefore, it is not clear how adequate the well-known solutions of the classical plane theory of elasticity are when applied to the analysis of actual plate components having finite thickness and what limitations have to be imposed on the use of these plane solutions.

An initial study of near crack tip fields accounting for effects of plate thickness was carried out by Hartranft and Sih [1]. They proposed an approximate three-dimensional theory and explored the influence of the plate thickness on the stress intensity factor. Kwon and Sun [2] have provided a comprehensive literature review on the earlier investigations of three-dimensional crack problems. A number of authors have reported that plane strain conditions prevail near the crack tip and that the plane strain solutions are only a good approximation of the three-dimensional stress field for very thick plates. Previous numerical and experimental studies have suggested that the three-dimensional stress region propagates from the crack front in the plane direction to a distance equal to approximately a half of the plate thickness. It was also suggested that the three-dimensional stress-strain behavior near the crack tip region is responsible for many experimentally observed fatigue and fracture phenomena, which can not be explained within plane stress or plane strain theories of elasticity [3, 4].

The mechanical argument behind the plate thickness effect on fracture initiation and fatigue crack growth rate is that the stress and strain fields in the vicinity of the crack tip are strongly dependent upon the specimen thickness. As the thickness increases the stress state will change from being largely plane stress to being plane strain dominant. This also has an effect on the size of the plastic zone at the crack tip, which will be greater for thin plates. The thickness effect is not independent and is a non-linear function of the loading, specimen geometry and material properties. Some of these effects will be addressed in the present work.

In the analysis of fracture problems in plate components, we adopt so-called, refined plate theory. The refined plate theory, which is also known as generalized plane strain theory or first order plate theory, was first introduced by Kane and Mindlin in 1956 in their work on high frequency extensional vibrations of circular disks [5]. The governing equations derived in this work include the transverse stress components while retaining the simplicity of a two-dimensional model.

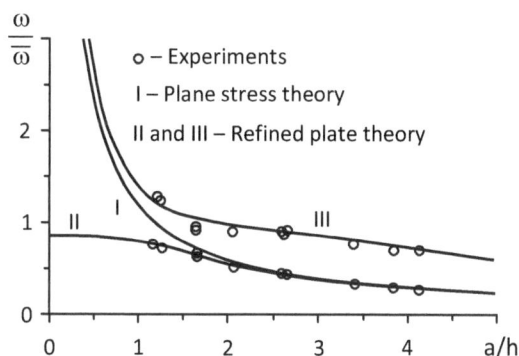

Figure 1: Comparison of theoretical predictions and experimental values for the natural frequencies of extensional vibrations of short circular disks [5].

A summary of the theoretical results obtained in this work is shown in Fig. **1** demonstrating that an application of refined plate theory provides a much better correlation with experimental data than the classical plane stress solutions. The significance of this theory is due to the fact that it is, probably, the only possible elementary extension of the classical plane theory of elasticity within which analytical results for non-trivial geometries (e.g. crack and notch problems) can be obtained and utilised for the practical analysis of three-dimensional stress, strain and displacement fields in plate components [3, 4].

GOVERNING EQUATIONS OF REFINED PLATE THEORY

In this section we describe basic equations of refined plate theory. Refined plate theory represents an elementary extension of the classical plane stress and plane strain theories of elasticity and mirrors these theories for very thin and very thick plates, respectively. Therefore, many classical results can be recovered from this theory as particular cases of more general solutions taking into account the plate thickness effect.

Variational Equation

Consider a homogeneous, elastic, and isotropic plate with a uniform thickness 2h, as shown in Fig. **2**. The mid-plane is chosen to be the x-y plane, with boundary planes at $z = \pm h$. For simplicity, the plan area A of the plate is taken to be simply connected. The plate has a right cylindrical boundary surface S, which intersects the middle plane along a closed contour C. The plate is in equilibrium under influence of the body force f_i over the volume V of the plate, and the surface traction p_i over a part of the surface of the plate where index (i) = (x, y, z). The body force, surface tractions and displacement boundary conditions are distributed symmetrically with respect to the mid-plane in such way that they do not generate bending deformations. In the refined plate theory the three-dimensional displacement field corresponding to the extensional deformations of the plate are approximated by the following equations [5]

$$u_x = u_x(x, y) \tag{1a}$$

$$u_y = u_y(x, y) \tag{1b}$$

and

$$u_z = k_s \frac{z}{h} w(x, y) \tag{1c}$$

where k_s is the shear factor. Kane and Mindlin suggest a value of $2\sqrt{3}/\pi$ for the shear factor in order to match the frequency of the lowest simple thickness-shear mode of vibration of an infinite plate given by the refined plate equations with that given by the exact elastic theory. However, in the present work the shear factor is chosen to be unity and will be omitted in the following considerations. This is justified by the fact that the theory under consideration at $h \to 0$ with $k_s = 1$ recovers the static plane-stress theory of elasticity [6].

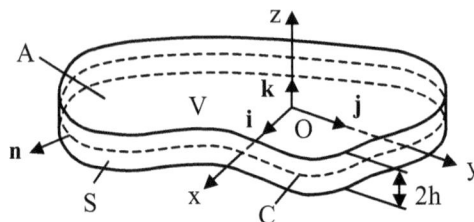

Figure 2: A plate.

By substituting Eqs (1) into the classical variational equation of the theory of elasticity by Timoshenko and Woinowsky-Krieger [7] and carrying out integration with respect to z over the thickness of the plate, 2h, we find [8]

$$\int_A \left(N_{\alpha\beta,\alpha} + P_\alpha^{(0)} + f_\alpha^{(0)} \right) \delta u_\alpha dA \; -\int_A \left(R_{a,\alpha} - N_{zz} + P_z^{(1)} + f_z^{(1)} \right) \delta w \, dA \; -\oint_C \left[\left(N_{\alpha\beta} n_\beta - p_\alpha^{(0)} \right) \delta u_\alpha \; + \left(R_\alpha n_\alpha - p_z^{(1)} \right) \right] \; \delta w dS = 0 \qquad (2)$$

Here n_α are the in-plane components of the normal unit vector, **n**, to the cylindrical boundary S (see Fig.2), the Greek indices $(\alpha, \beta) = (x, y)$, the summation rule applies for repeated indices, and a comma means the differentiation, or $(\cdot)_{,\alpha} = \dfrac{\partial(\cdot)}{\partial\alpha}$. In the previous variational equation

$$p_\alpha^{(0)} = \int_{-h}^{h} p_\alpha dz \qquad (3a)$$

$$P_\alpha^{(0)} = p_\alpha^+ + p_\alpha^- \qquad (3b)$$

$$P_z^{(1)} = h(p_z^+ + p_z^-) \qquad (3c)$$

are the plate surface tractions and the superscripts (0) and (1) refer to quantities of the zeroth and first order, respectively, in the thickness coordinate, $p_i^\pm$ denote surface tractions at $z = \pm h$, and because of the symmetry of loading with respect to the mid-plane $p_i^+ = p_i^- = p_i$.

The plate body forces are

$$f_\alpha^{(0)} = \int_{-h}^{h} f_\alpha dz \qquad (3d)$$

$$f_z^{(1)} = \int_{-h}^{h} f_z z \, dz \qquad (3e)$$

The in-plane stress resultants are defined as

$$N_{\alpha\beta} = \int_{-h}^{h} \sigma_{\alpha\beta} dz \qquad (3f)$$

Thus, $N_{\alpha\beta}$ is equal to the average through-the thickness in-plane stress, $\sigma_{\alpha\beta}$, multiplied by the plate thickness.

The out-of-plane resultants are

$$R_\alpha = \int_{-h}^{h} \sigma_{z\alpha} z \, dz \tag{3g}$$

$$N_{zz} = \int_{-h}^{h} \sigma_{zz} dz \tag{3h}$$

Therefore, N_{zz} is equal to the average transverse normal stress multiplied by the plate thickness, and R_α are the components of pinching shear, a term used by Kane and Mindlin. The latter play a role in extension deformations similar to the transverse shearing forces in flexure. The variational equation (2) is the two-dimensional equation that can be rewritten in the differential form [8]

$$N_{\alpha\beta,\beta} + P_\alpha^{(0)} + f_\alpha^{(0)} = 0 \tag{4a}$$

$$R_{\alpha,\alpha} - N_{zz} + P_z^{(1)} + f_z^{(1)} = 0 \tag{4b}$$

with the boundary conditions on C obtained from Equation (2) in the following form

$$N_{\alpha\beta} n_\beta - p_\alpha^{(0)} = 0 \ \text{ or } \ u_\alpha = \overline{u}_\alpha \tag{5a}$$

$$R_\alpha n_\alpha - p_z^{(1)} = 0 \ \text{ or } \ w = \overline{w} \tag{5b}$$

where $\overline{u}_\alpha$ and $\overline{w}$ are prescribed functions (boundary conditions) on contour C.

Note that, by setting $w = 0$ in Eq. (2) the classical plane strain theory of extension can be recovered.

Displacements, Strains and Stress Resultants

Taking into account the kinematic assumption of the refined plate theory, Eq. (1c), the strain-displacement relations are:

$$\varepsilon_{\alpha\beta} = \frac{u_{,\alpha} + u_{,\beta}}{2} \tag{6a}$$

for the in-plane strain components, and

$$\varepsilon_{zz} = \frac{w}{h} \tag{6b}$$

$$\varepsilon_{z\alpha} = z \frac{w_{,\alpha}}{2h} \tag{6c}$$

for the out-of-plane strain components.

Substitution of Eqs (6) into Hooke's law yields the following stress resultant - displacement relationships:

$$\frac{N_{\alpha\beta}}{2h} = \lambda\theta\delta_{\alpha\beta} + \mu\left(u_{\alpha,\beta} + u_{\beta,\alpha}\right),$$ (7a)

$$\frac{N_{zz}}{2h} = \lambda\theta + 2\mu\frac{w}{h}$$ (7b)

$$\frac{R_\alpha}{2h} = \frac{h}{3}\mu w_{,\alpha}$$ (7c)

The dilatation $\theta = u_{\alpha,\alpha} + \dfrac{w}{h}$, and the Lame constants are λ and μ.

Strain Compatibility Equation

Consider an elastic isotropic plate as shown in Fig. **2** being in equilibrium. In the absence of body forces and surface tractions the equations of equilibrium (4) can also be rewritten as [9]

$$N_{\alpha\beta,\beta} = 0$$ (8a)

$$\nabla^2 w - \frac{6(1+\nu)}{h^2}w = \frac{6\nu(1+\nu)}{h^2 E}N$$ (8b)

where $N = N_{\alpha\alpha}/2$ is the mean in-plane stress resultant, $\nabla^2 w = w_{,\alpha\alpha}$ is the two dimensional Laplace operator (where the summation rule applies for repeated indices), E is Young's modulus and ν is Poisson's ratio.

For a particular problem, it is necessary to the solve differential equations of equilibrium, and the solution must be such that it satisfies the boundary conditions. These differential equations containing three stress resultant components $N_{\alpha\beta}$ and out-of-plane displacement component w are not sufficient for the determination of these components. The problem is a statically indeterminate one, and in order to obtain the solution the additional condition of compatibility must also be considered. In rectangular coordinates it takes the following form [10]

$$\varepsilon_{x,xx} + \varepsilon_{y,yy} = 2\varepsilon_{xy,xy}$$ (9)

Using strain-stress resultants relationships (7) and equilibrium equations (8), the compatibility equation (9) can be written as [6, 9]

$$\nabla^2 N - \frac{\nu}{2}\nabla^2 N_{zz} = 0$$ (10)

The equations of equilibrium (8) together with the boundary conditions (5) and compatibility equation (10) form a complete system of equations that is usually sufficient for the complete determination of all stress resultants and out-of-plane displacement function, w.

LINEAR-ELASTIC FRACTURE MECHANICS

In this section we consider a through-the-thickness crack in an infinite elastic plate of finite thickness, which is stressed in mode I (opening mode of fracture). We will apply fundamental solution for an edge dislocation obtained in the work of Kotousov and Wang [11] to analyse the three-dimensional stress and displacement fields. The classical plane theories of elasticity, utilising plane stress or plane strain assumptions, predicts either zero or infinite out-of-plane displacements (u_z) at the crack tip, respectively. However, numerous experimental and numerical results as well as common sense, suggest that the u_z-displacement at the crack tip is finite and not equal to zero. This fundamental discrepancy, in particular, indicates that neither plane stress or plane strain conditions are a valid assumption to analyse the three-dimensional stress and displacement fields near the tip of the through-the-thickness crack.

The computational approach adopted in this and the following sections is based on the distributed dislocation theory, which was established by Liebfried in early 1950's [12], although then the target was physical dislocations with discrete Burgers vector and discrete position. The mathematical concept of distributed dislocation arrays with continuos Burgers vectors is based on the pioneering work of Eshelby in the late 1950's [13], and developed by Erdogan [14], Keer, Mura [15] and many others. The basic idea is to use the superposition principle for the uncracked body, together with an unknown distribution of dislocations placed along the crack, chosen so that the crack faces become traction-free. The technique is very efficient, and may be applied to model cracks in two and three dimensions. The techniques may be used to determine very accurately the stress intensity factor as well as the crack opening displacement.

Semi-Infinite Crack in a Finite Thickness Plate

We start with consideration of a semi-infinite crack. It represents a very important idealization in fracture mechanics as an application of this idealization normally leads to the reduction in the number of parameters controlling crack tip behaviour and fracture. This section will begin with an outline of the distributed dislocation technique as applied to a straight semi-infinite crack in a plate of thickness 2h. A full description of the technique can be found in Codrington and Kotousov [16], however for completeness a brief review will be presented here.

It is assumed that a through-the-thickness crack lies along the x-axis, as shown in Fig. **3**, $(-\infty < x < 1)$ in an infinite plane and is subjected to a remotely applied mode I stress intensity factor, K_I^{app}. By replacing the crack with a continuous distribution of dislocations along the x-axis and by use of the superposition principle, the problem is reduced to the following singular integral equation [17]:

$$\sigma_{yy}(x) = \frac{1}{\pi} \int_{-\infty}^{1} b_y(\xi) g_{yy}(x;\xi) d\xi \tag{11}$$

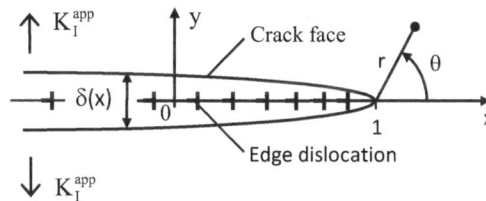

Figure 3: A semi-infinite crack in plate represented by continuously distributed dislocations.

In the previous equation $b_y(\xi)$ is the unknown dislocation density function and is related to the separation of the crack faces, $\delta(\xi)$, by:

$$b_y(\xi) = -\frac{d\delta(\xi)}{d\xi} \tag{12}$$

The function $g_{yy}(x; \xi)$ is the dislocation influence function, which forms the singular kernel of the system. This function represents the non-dimensional stress at a point x (y = 0) due to a dislocation with a unit Burgers vector located at point ξ on the x axis. In the case of plane stress or plane strain the influence function is given as [17]:

$$g_{yy}(x;\xi) = \frac{2\mu}{k+1} \frac{1}{x-\xi} \tag{13}$$

where μ is the shear modulus, k is Kolosov's constant (equal to $(3 - v)/(1 + v)$ in plane stress and $3 - 4v$ in plane strain with v being Poisson's ratio).

The three-dimensional formulation of this problem utilizes the refined plate theory and can be obtained by replacing the two-dimensional kernel g_{yy} with its three-dimensional analog, G_{yy}. The three-dimensional kernel is obtained from the solution for an edge dislocation in an infinite plate of finite thickness, which was derived by Kotousov and Wang [11] as:

$$G_{yy}(x;\xi) = -\frac{E}{4(1-v^2)}\frac{1}{x-\xi}\left[\frac{4v^2}{(\kappa\rho)^2} - (1-v^2) - 2v^2 K_0(\kappa\rho) - 2\frac{(2+\kappa^2\rho^2)v^2 K_1(\kappa\rho)}{\kappa\rho}\right] \tag{14}$$

where $K_n(\cdot)$ are the modified Bessel functions of nth order and the parameter κ is given by:

$$\kappa = \frac{1}{h}\sqrt{\frac{6}{1-v}} \tag{15}$$

and:

$$\rho = |x-\xi| \tag{16}$$

The second boundary condition on the crack faces, which is the absence of the transverse shear stress or stress resultant, $R_y = 0$, is also satisfied due to the dislocation solution [11].

To solve the integral equation (11) with kernel (14) via Gauss-Chebyshev (GC) quadrature [18] we first introduce the coordinate transformations:

$$x = \frac{2t}{t+1} \text{ and } \xi = \frac{2s}{s+1}, \tag{17}$$

such that the transformed integral equation becomes:

$$\frac{N_{yy}(x)}{2h} = \frac{1}{\pi}\int_{-1}^{+1} B_y(s)G_{yy}(x;s)\frac{2}{(s+1)^2}\,ds \tag{18}$$

where $B_y(s)$ is the dislocation density corresponding to the three-dimensional formulation.

Solution to the singular equation (18) is obtained by applying GC quadrature to form a system of N - 1 linear equations with N unknowns:

$$\frac{N_{yy}(x_j)}{2h} = \frac{1}{n}\sum_{i=1}^{n}\phi(s_i)G_{yy}(x_j;s_i)\frac{2}{s_i+1} \tag{19}$$

where $j = 1... n - 1$ and n is the number of integration points and the non-singular function $\phi(s)$ is related to the dislocation density by:

$$\frac{B_y(s)}{1+s} = \phi(s)(1+s)^{-1/2}(1-s)^{-1/2} \tag{20}$$

The discrete integration and collocation points are given, respectively, as:

$$s_i = \cos\left[\pi\frac{2i-1}{2n}\right] \quad i = 1... n, \tag{21}$$

and:

$$t_j = \cos\left[\pi\frac{k}{n}\right] \qquad j = 1 \dots n - 1. \tag{22}$$

The through-the-thickness averaged crack tip stress intensity factor can be determined from an asymptotic analysis of the crack tip opening displacement as:

$$K_I = \lim_{r \to 0} \sqrt{2\pi r}\, \frac{2\mu}{\kappa + 1} \frac{d\delta(r)}{dr} = \sqrt{2\pi}\, \frac{2\mu}{\kappa + 1} \phi(1) \tag{23}$$

in the case of plane stress or plane strain and:

$$K_I = \sqrt{2\pi}\, \frac{E}{4(1 - v^2)} \phi(1) \tag{24}$$

for the finite thickness plate. The approximation of $\phi(1) = \phi(s_1)$ is usually made. From the above equations it is seen that the asymptotic analysis of the finite thickness plate (24) has resulted in the plane strain form of (23).

The nth equation required to solve the set of linear equations (19) is given by the condition of the remotely applied stress intensity factor such that:

$$\phi(s_n) = \frac{K_I^{app}}{\sqrt{2\pi}} \frac{k + 1}{2\mu} \tag{25}$$

for plane stress or plane strain and:

$$\phi(s_n) = \frac{K_I^{app}}{\sqrt{2\pi}} \frac{4}{E} \tag{26}$$

for the finite thickness plate. It is assumed in the finite thickness problem that plane stress conditions will prevail remote from the crack tip [9] and therefore Eq. (26) is taken as the plane stress form of (25).

The requirement that the crack faces remain traction free provides the additional condition that $N_{yy}(x) = 0$ along the line of the crack. The set of n linear equations and n unknowns, $\phi(s_i)$ for i = 1 n, can now be solved via any standard method.

Let us now consider selected results for the out-of-plane (or transverse) stress field near crack tip of a semi-infinite crack in a plate of finite thickness. The effect of the normal transverse stress component is commonly described in the literature by the out-of-plane constraint factor, which can be defined as [6]:

$$T_z(r, \theta) = T_z(x, y) = \frac{1}{v} \frac{N_{zz}(x, y)}{N_{xx}(x, y) + N_{yy}(x, y)} \tag{27}$$

The constraint factor has a simple interpretation: $T_z(x, y) \to 0$ means the plane stress conditions prevail; and $T_z(x, y) \to 1$ suggests that the plane strain state is dominant at this location. Stress resultants $N_{xx}(x, y)$, $N_{yy}(x, y)$ and $N_{zz}(x, y)$ are found in a similar manner to Eqs (14) and (18). For simplicity, the kernel of the stress sum, $N_{xx}(x, y) + N_{yy}(x, y)$, is given by:

$$G_{xx+yy}(x, y; \xi) = \frac{E}{2(1 - v^2)} \frac{x - \xi}{\rho^2}\left[(1 - v^2) + \frac{\kappa^2 \rho^2 v^2 K_1(\kappa\rho)}{\kappa\rho}\right] \tag{28}$$

and for $N_{zz}(x, y)$ is:

$$G_{zz}(x,y;\xi) = \frac{E}{2(1-v^2)} \kappa v K_1(\kappa\rho)\frac{x-\xi}{\rho} \qquad (29)$$

where now:

$$\rho^2 = (x-\xi)^2 + y^2 \qquad (30)$$

These kernels are obtained from the fundamental solution for an edge dislocation in infinite plate similar to Eq. (14) [11]. The conversion between Cartesian and polar coordinates can be made via the usual transformations (see Fig. 3):

$$x = r\cos\theta + 1 \text{ and } y = r\sin\theta \qquad (31)$$

Results for the mid-thickness out-of-plane constraint factor are shown in Fig. 4 as a function of the radial distance from the crack tip to plate thickness ratio ($\theta = 0°$). FE results by Nakamura and Parks [19], and She and Guo [20] for semi-infinite cracks are also given as a comparison and are in a good agreement with the present results. It can be seen that at the crack tip the conditions reach near plane strain state, while at approximately $r/2h \sim 1.5$ the plane stress solution is recovered. In these calculations Poisson's ratio was taken as $v = 0.3$, although any variation of v has minimal effect on the constraint factor. Approximately 250 integration points were required to reach a convergence in the solution. Fig. 5 shows the results for the out-of-plane constraint factor as a function of θ for various radius to plate thickness ratios. FE results by [20] are given and show a good agreement with the present results. The crack tip stress intensity factor for the finite thickness pate, in the case of a semi-infinite crack, can be determined using Eq. (24). Results for the ratio of the crack tip stress intensity factor to the far-field stress intensity factor (the remotely applied stress intensity factor) are shown in Fig. 6 as a function of Poisson's ratio. As it can bee seen from this figure, the present results obtained utilising the distribution technique and refined plate theory are identical to the through-the-thickness average values presented by She and Guo [20], Nakamura and Parks [19], and Yang and Freund [9] for semi-infinite cracks.

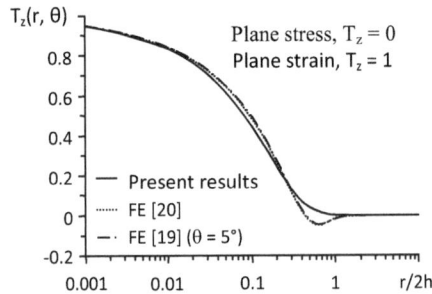

Figure 4: Mid-thickness out-of-plane constraint factor as a function of the radial distance from the crack tip to plate thickness ratio ($\theta = 0°$ and $\theta = 5°$).

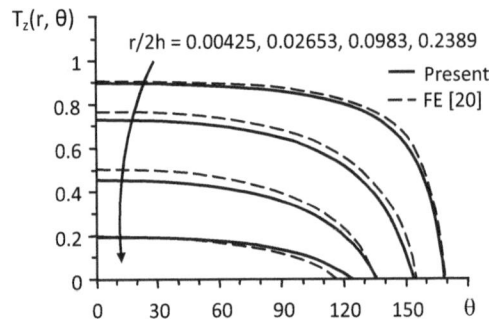

Figure 5: Mid-thickness out-of-plane constraint factor as a function of θ for various radial distances from the crack tip to plate thickness ratios.

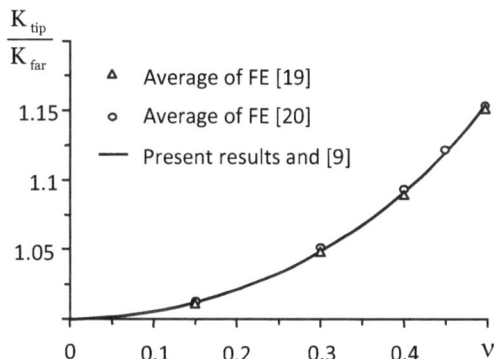

Figure 6: Ratio of the crack tip stress intensity factor to far-filed stress intensity factor as a function of Poisson's ratio for a semi-infinite crack.

Using the superposition principle and dislocation solution (18) – (22), the out-of-plane displacement, u_z, can be determined at any point within the plate as:

$$u_z(r,\theta,z) = u_z(x,y,z) = \frac{z}{h}\frac{1}{\pi}\int_{-\infty}^{1} B_y(\xi)G_{uz}(x,y;\xi)d\xi \qquad (32)$$

where the $z = 0$ at the plate mid-thickness and $z = \pm h$ on the plate surfaces. The displacement kernel for the finite thickness plate is given by:

$$G_{uz}(x,y;\xi) = -\frac{\kappa v h}{2}\frac{x-\xi}{\rho}\left[\frac{1}{\kappa\rho} - K_1(\kappa\rho)\right] \qquad (33)$$

which can be obtained from the three-dimensional fundamental solution for an edge dislocation in infinite plate [11]. The displacement kernels for classical plane theory of elasticity are:

$$g_{uz}(x,y;\xi) = -\frac{v h}{2}\frac{x-\xi}{\rho^2} \qquad (34a)$$

for plane stress:

$$g_{uz}(x,y;\xi) = 0 \qquad (34b)$$

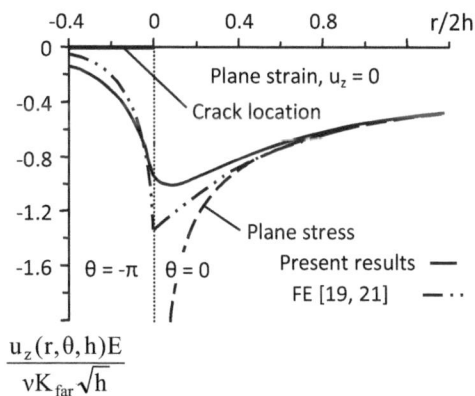

Figure 7: Normalized out-of-plane surface displacement as a function of the radial distance from the crack tip to plate thickness ratio in the case of a semi-infinite crack.

for the plane strain assumption. It can be shown that the kernel given by Eq. (33) recovers correctly the results of plane theory of elasticity as limiting cases of very thin and very thick plates, respectively.

Selected results for normalized out-of-plane surface displacement along the line of the crack are presented in Fig. 7 for the case of a semi-infinite crack. Pfaff et al.'s [21] empirical fit to the FE results by Nakamura and Parks [19] is also given as a comparison. Some difference between the finite element results and the theoretical predictions could be explained by mesh refinement issues at the crack tip in the FE model and finite geometries used in FE simulations as well as the underlying kinematic assumption of the refined plate theory.

Finite Crack in an Infinite Plate

The formulation of the finite length crack problem is very similar to that of the semi-infinite crack as outlined above. It is assumed that a through-the-thickness crack of length 2a lies within -a < x < a on the x-axis in an infinite plane and is subjected to remotely applied stress tractions, σ_{yy}^{app} (Figs. **8** and **9**).

The solutions of the plane stress and refined plate theories are the same far from the crack subjected to uniaxial loading perpendicular to the crack faces i.e. $\sigma_{yy}^{app} = N_{yy}^{app}/2h$. The governing singular integral equation for this problem therefore becomes [17]:

$$\frac{N_{yy}(x)}{2h} = \frac{1}{\pi} \int_{-a}^{a} B_y(\xi) G_{yy}(x;\xi) d\xi + \sigma_{yy}^{app}$$

(35a)

In addition, the dislocation density function, B_y, also satisfies the following single-valuedness condition:

$$\int_{-a}^{a} B_y(\xi) d\xi = 0$$

(35b)

In other words, this means that moving from one end of the crack to the other there should be no net dislocation, or $\delta(-a) = \delta(a) = 0$.

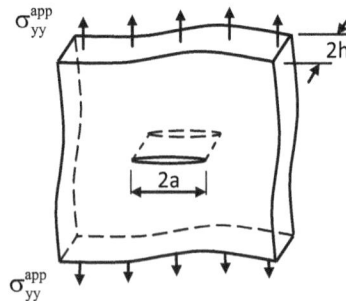

Figure 8: Crack in plate of finite thickness, 2h, subjected to remote loading.

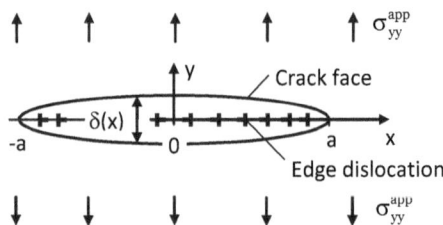

Figure 9: Crack in plate represented by continuously distributed dislocations.

Solution to the integral equation (35a) with condition (35b) follows via application of GC quadrature in a similar manner as for the semi-infinite crack case. The transformations (17), however, are replaced with the new transformations [17]:

$$x = at \text{ and } \xi = as \tag{36}$$

and such that for along the crack $|t|$ and $|\xi| < 1$ and the non-singular function $\phi(s)$ is now related to the dislocation density by:

$$B_y(s) = \phi(s)(1+s)^{-1/2}(1-s)^{-1/2} \tag{37}$$

This leads to the following system of algebraic equations:

$$\frac{a}{n}\sum_{i=1}^{n}\phi(s_i)G_{yy}(x_j;s_i)+\sigma_{yy}^{app}=0 \tag{38a}$$

$$\frac{\pi}{n}\sum_{i=1}^{n}\phi(s_i)=0 \tag{38b}$$

where the discrete integration points are given by:

$$s_i = \cos\left[\pi\frac{2i-1}{2n}\right], \quad i = 1...n, \tag{39a}$$

$$t_j = \cos\left[\pi\frac{j}{n}\right], \quad j = 1...n-1. \tag{39b}$$

Equations (38) represent a system of n algebraic equations to the n unknowns $\phi(s_i)$, $i = 1...n$. The averaged stress intensity factor at the two crack tips are given by:

$$K_I = \frac{E\sqrt{\pi a}}{4(1-v^2)}\phi(\pm 1) \tag{40}$$

which is similar to the case of a semi-infinite crack considered previously, see Eq. (24).

Results of numerical solution are shown in Fig. **10** in terms of the stress-intensity magnification factor, which is defined as a ratio of the averaged stress intensity factor to its plane stress counterpart:

$$F = \frac{K_I}{\sigma_{yy}^{app}\sqrt{\pi a}} \tag{41}$$

as a function of the ratio of the crack length to the plate thickness. The results demonstrate that although the stress-state at the vicinity of a crack tip is always plane-strain, the magnitude of the stress-intensity factor depends on the ratio of crack size to plate thickness. In particular, in the short crack limit or very thick plate, i.e. $a/h \to 0$, the stress intensity magnification factor F = 1, suggesting the validity of the classical plane solution

$$K_I(a/h \to 0) = \sigma_{yy}^{app}\sqrt{\pi a} \tag{42}$$

In the case of a long crack (i.e. $a/h \to \infty$) the magnification factor is related to the classical plane solution by $1/\sqrt{1-v^2}$. This result has been obtained earlier for semi-infinite crack [9] and may be verified by direct application of Rice's path-independent J-integral [22]. From Fig.10 is also seen that the transition from plane strain to plane stress conditions occurs within the region $0.1 < a/h < 10$.

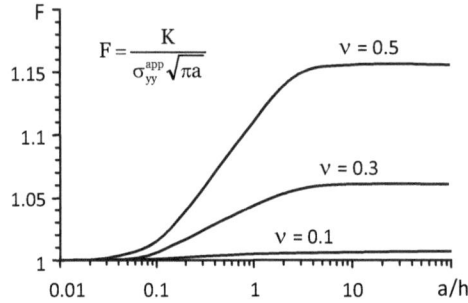

Figure 10: Stress intensity magnification factor for embedded crack in a finite thick plate as a function of the ratio of the crack length to the plate thickness.

The out-of-plane displacement for the finite length crack case may be determined by:

$$u_z(r,\theta,z) = u_z(x,y,z)$$
$$= \frac{z}{h}\frac{1}{\pi}\int_{-a}^{a} B_y(\xi) G_{uz}(x,y;\xi)d\xi - \frac{z\nu}{E}\sigma_{yy}^{app} \qquad (43)$$

where the extra term in the equation is due to uniform lateral contraction of the infinite plate loaded by the remote tensile stress σ_{yy}^{app}.

Fig. **11** displays the results for the normalized out-of-plane displacement ahead of the crack tip for the case of a finite length crack. The present results are in good agreement with the experimental values of Humbert *et al.* [23]. There are some differences between the results near the crack tip, although it is less than 10% and can be partially explained by the presence of the process zone where the material is subjected to non-elastic deformations.

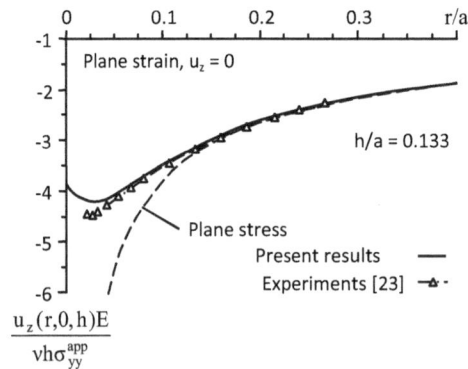

Figure 11: Normalized out-of-plane surface displacement as a function of the radial distance from the crack tip to plate thickness ratio ($\theta = 0°$).

NON-LINEAR FRACTURE MECHANICS

This section will begin with a generalization of the classical strip-yield model, or Dugdale model [24], which was originally suggested for very thin plates. This generalization will extend the simple 2D model for use with a through-the-thickness crack in a plate of finite thickness and stressed in mode I. For this purpose we introduce some assumptions and simplifications, which will enable analytical results to be obtained based on the refined plate theory. Then, we briefly describe the plasticity-induced crack closure concept and develop computational techniques for calculating its effect on the fatigue crack growth rate under both constant and variable amplitude loading. As a result, a number of very important non-liner crack growth phenomena can be investigated accurately and efficiently without use of extensive, time-consuming and, sometimes, controversial finite element simulations. However, the range of applicability of the results for a growing crack is limited to materials and load conditions where the plasticity-induced crack closure mechanism is dominant.

A Generalized Strip-Yield Model

Similar to the original model, plasticity is assumed to occur within a strip directly ahead of the crack tip [25]. Furthermore, the plastic zone size is assumed to be constant in the thickness direction, thus averaging the plastic zone through the plate thickness. In order to obtain analytical results, both the crack and the plastic zones are represented by a distribution of displacement discontinuities, or edge dislocations.

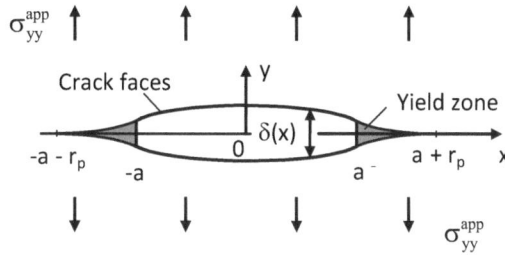

Figure 12: Strip-yield model.

Adopting a Tresca yield criterion for stress resultants, the yielding condition within the plastic zone can be written as,

$$\left| \frac{N_{yy}}{2h} - \frac{N_{zz}}{2h} \right| = \sigma_f \tag{44}$$

where σ_f denotes the material's flow stress (normally the average between the yield stress, σ_y, and ultimate stress, σ_{ult}). In plasticity criterion (44) the condition that N_{xx} is less than N_{yy} has been used; on solving the problem, it turns out that this is indeed the case. Thus, the equilibrium condition can be written:

for $|x| < a$

$$\frac{1}{\pi} \int_{-a-r_p}^{a+r_p} B_y(\xi) G_{yy}(x;\xi) d\xi + \sigma_{yy}^{app} = 0 \tag{45a}$$

and for $a \le |x| \le a + r_p$

$$\frac{1}{\pi} \left| \int_{a-r_p}^{a+r_p} B_y(\xi)\big(G_{yy}(x;\xi) - G_{zz}(x;\xi)\big) d\xi \right| + \sigma_{yy}^{app} = \sigma_f \tag{45b}$$

where the kernel G_{yy} is given by (14), and G_{zz} by, referring to Eq. (29).

The plastic zone size r_p is determined using the fact that the stress at the tip of the plastic zone must be finite. This can be expressed mathematically as:

$$K_I(a + r_p) = K_I(-a - r_p) = 0 \tag{46}$$

Solution of the plastic zone size r_p can be obtained iteratively by solving Eq. (45) and calculating the stress intensity factor at the end of the plastic zone until the required accuracy of Eq. (46) is reached. The results are shown in

Fig.13 for three different ratios of crack size to plate thickness, together with the solutions for plane-stress and plane-strain. An explicit solution for the plastic zone size can be derived for the plane-strain and plane-stress problems, such as the one given by Kelly and Nowell [26]. For the plane strain it was shown that:

$$\frac{a}{a+r_p} = \cos\left[\frac{\pi}{2}\frac{(1-2\nu)\eta}{1-\nu\eta}\right], \quad \eta = \frac{\sigma_{yy}^{app}}{\sigma_f} \tag{47}$$

The plane stress solution is recovered by setting $\nu = 0$.

It is clear that for low applied stress (relative to the material's flow stress) the present solutions do indeed recover the plane-strain solutions. However, at high applied stress ratio ($\sigma_{yy}^{app}/\sigma_f \to 1$), the plastic zone size approaches the plane-stress solution (see Fig. **13**). Furthermore, in addition to the applied stress, the ratio of crack-size to plate thickness also has a significant effect on the extent of the plastic zone, r_p.

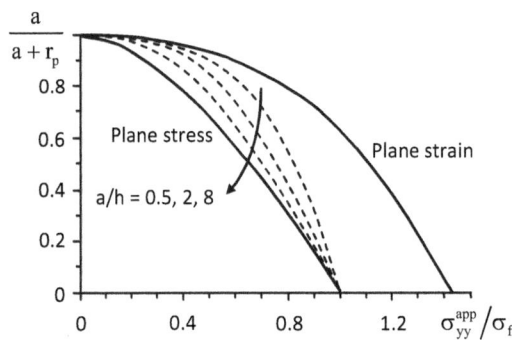

Figure 13: Plastic zone size as a function of the ratio of the applied load to yield load.

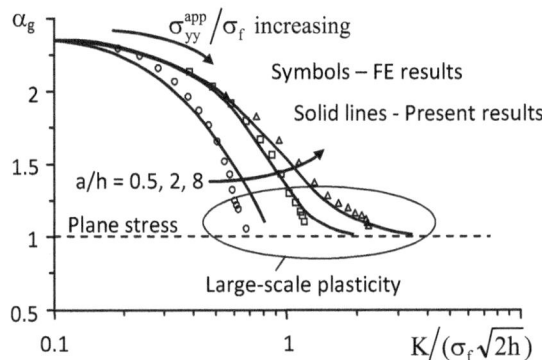

Figure 14: Theoretical and FE [27] predictions of the global constraint factor as a function of non-dimensional stress intensity factor for various crack length to the plate thickness ratios.

A comparison of the generalized strip-yield model with finite element results can be made via the, so called, plastic constraint factor, α_g. The plastic constraint factor has been successfully employed to simulate the effect of three-dimensional constraint on fatigue crack growth in plates of arbitrary thickness in many previous works and computer-based programs. In accordance with Newman et al, [27], it can be defined as:

$$\alpha_g = \frac{1}{\sigma_f r_p}\int_a^{a+r_p}\sigma_{yy}(x)dx \tag{48}$$

Where:

$$\sigma_{yy}(x) = \frac{N_{yy}(x)}{2h} = \sigma_{yy}^{app} + \frac{1}{\pi} \int_{-a-r_p}^{a+r_p} B_y(\xi)G_{yy}(x;\xi)d\xi \qquad\qquad (49)$$

The obtained theoretical dependencies for the constraint factor are shown in Fig. **14** together with the numerical results by Newman et al [27] who used the three-dimensional elasto-plastic finite element method. From this figure it is seen that there is good overall correlation between the present theory and the FE results. In the large-scale plasticity region the results diverge as the numerical model used by Newman et al [27] experiences the finite width effect with the plastic zone approaching the model boundaries.

A Generalized Crack Closure Model for Fatigue Crack Growth

Most engineering components or structures experience some form of variable loading events, such as overloads or underloads, throughout their service lifetime. The crack growth rate for a given material can vary significantly with the history of loading, applied load ratio as well as the thickness of the component or specimen. It is therefore crucial that these factors are accurately accounted for when making fatigue crack growth predictions and fatigue lifetime assessments.

The discovery of fatigue crack closure by Elber in late 60s and early 70s [28, 29] provided a new insight into the mechanisms of fatigue growth. Plasticity-induced crack closure (PICC) was first introduced by Elber after he observed that fatigue cracks can remain closed even when the applied load is tensile. Elber attributed the closure to residual plastic deformation remaining on the crack faces as the crack propagates. The crack closure concept, in particular plasticity-induced crack closure, has been successfully used to explain a number of fatigue phenomena. These include the effects of load ratio, retardation due to an overload cycle, short crack growth, and specimen thickness (see Newman [30] and Skorupa [31]). Plasticity-induced crack closure has been observed in many ductile metals and alloys including, but not limited to, steel (Dougherty [32], Skorupa and Skorupa [33]), aluminium alloys (Elber [28, 29], Schijve [34], McEvily and Ishihara [35]), and titanium alloys (Bachmann and Munz [36]), as well as in polymers such as polycarbonate (James et al [37]). Through the use of plasticity-induced crack closure it is also possible to collapse fatigue crack growth rate curves obtained for different applied load ratios or plate thicknesses, for a given material, onto a single unique baseline curve (e.g. Elber [29], Codrington and Kotousov [38]).

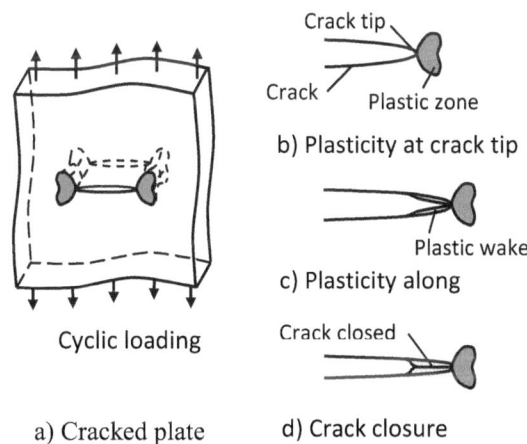

Figure 15: Plasticity-induced fatigue crack closure.

The underlying principle of PICC is that as a fatigue crack propagates a region of plastic deformation is formed at the crack tip (Fig. **15a,b**). This plastically deformed material then gets left along the crack faces as a plastic wake as the crack continues to grow (Fig. **15c**). The extra layer of material along the crack faces leads to premature closure during the unloading portion of the load cycle (Fig.15d). It was proposed by Elber [28] that fatigue damage is minimal when the crack is closed and that significant crack growth will only occur when the crack is fully open. He then suggested the use of an effective stress intensity factor range:

$$\Delta K_{eff} = K_{max} - K_{op} \tag{50}$$

where K_{max} is the maximum stress intensity factor for a given load cycle, and K_{op} refers to the stress intensity factor at the point when the crack tip just re-opens. This effective range is used in place of the traditional linear elastic stress intensity factor range in theoretical fatigue life models as well as for correlating constant amplitude fatigue crack growth data. Various sources of closure have been identified since Elber's [28] discovery, including surface roughness and oxide debris (Suresh, [39]). Region I crack growth (near threshold) is thought to be controlled by debris and surface roughness closure (Blom & Holm [40]; Dougherty et al., [32]). However, within region II crack growth, or the 'Paris regime', PICC is usually found to be the dominant mechanism (Suresh, [39]; Dougherty *et al.*, [32]).

Many experimental methods have been utilised to investigate plasticity-induced closure including: compliance techniques, which involve near and far-field strain measurement using strain gauges or crack mouth displacement gauges; optical techniques like Moiré interferometry or video; the potential drop method and also acoustic emission. A comprehensive review of crack opening/closure load measurement is provided by Stoychev and Kujawski [41]. These studies have demonstrated that there are many factors that influence the growth of fatigue cracks Skorupa [42]). They have also shown that there can be significant interaction between the plate thickness, the applied loading history and the material properties (Skorupa, [31] and [42]). For example, the application of a tensile overload cycle can provide substantial crack growth retardation and even complete crack arrest (Shuter and Geary [43], Borrego et al., [44] and Skorupa [42] among many others). The extent of retardation has been found to be dependent not only on the overload and baseline loading, but also on the thickness of the component (Shuter and Geary [43] Borrego *et al.* [44]). In general, an increase in the plate thickness means a reduction in the extent of overload retardation.

A large number of prediction methods have been developed over the past 100 years, ranging from the traditional safe-life approaches (e.g. Basquin [45], Fatemi and Yang [46]) to crack propagation based theories (e.g. Paris and Erdogan [47]). In particular, the use of crack propagation theories and the concept of plasticity-induced crack closure have proven to be quite popular. The most commonly utilised approach is that of a plane stress (i.e. two dimensional) strip-yield model by Dugdale [24], which assumes all plastic deformation occurs in an infinitesimal strip along the line of the crack. Early work by Dill and Saff [48], Budiansky and Hutchinson [49] and Führing and Seeger [50] paved the way for a large number of researchers who have since embraced this simplified method. Newman [51] extended upon these methods by introducing a correction, or constraint, factor in an attempt to account for the three-dimensional geometry effects. Reasonable success has been achieved in comparison to experimental results (see Newman [51]); however, there is much difficulty in discerning an appropriate value for the constraint factor. Trial-and-error is often employed in order to determine a value, or sometimes a simple function, for the constraint factor. Curve fitting based on the correlation of constant amplitude fatigue test data may also be used (Newman et al., [52]). This greatly limits the ability of the models to accurately predict the interaction between the involved variables, for example the thickness and load history.

An alternative approach for investigating the thickness effect on crack closure and fatigue crack growth is through the use of finite element methods. Finite element (FE) methods have successfully been used to model complex two and three-dimensional geometry for a wide range of load conditions (Solanki *et al.*, [53]). While being able to give a better understanding of the crack closure mechanism, finite element methods suffer many numerical issues and can have large computational requirements. Such issues include mesh refinement and convergence, the crack advancement scheme, and crack face contact. These limitations make FE methods unsuitable for systematic investigation of a wide range of the input parameters and limit their use for practical fatigue life prediction. In the next section we introduce a semi-analytical method, which is based on the generalized Dugdale model and free from all these shortcomings. This method is shown to be an excellent alternative to finite element based techniques for the investigation of non-linear fatigue crack growth phenomena.

CALCULATION PROCEDURE

In order to generalise the results for constant amplitude loading, the conditions of small-scale yielding (SSY) are utilised. Under small-scale yielding conditions the effects of the in-plane plate geometry, such as specimen type, plate width and stress concentrators are negligible. This is due to the size of the tensile plastic zone being far less than any in-plane characteristic lengths including the crack length, remaining ligament size, etc.

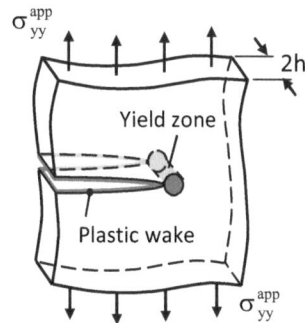

Figure 16: Schematic of a semi-infinite crack in an infinite plate

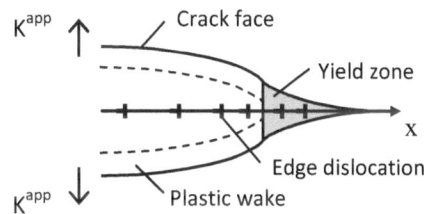

Figure 17: Schematic of dislocation representation.

Use of SSY conditions therefore allows for the generalisation of the analysis to a wide range of plate geometries. Small-scale yielding conditions are generally considered to be appropriate when the maximum applied stress to yield stress ratio is no greater than ~ 0.2-0.3. The SSY assumption makes it only necessary to consider a small region surrounding the crack tip and thus semi-infinite crack geometry is adopted (Fig. **16**). It is further assumed that the plastic wake is fully developed and that the crack has been growing for a significant period under constant amplitude loading. This allows for the plastic wake that is already developed along the crack length to be treated as a layer of constant thickness.

A rigid-perfect-plastic strip-yield model is again utilised for the regions of plastic deformation ahead of the crack tip and also for the plastic wake (Fig. **17**). The crack and plastic zones are then mathematically modelled by way of a distribution of edge dislocations. It is understood that the strip-yield hypothesis is most applicable to a plane stress analysis; however, it does provide a suitable approximation and modelling simplification for the three-dimensional case (Newman [51], Wang and Blom [54], Kotousov and Wang [11] and Codrington [55]). It should be noted that Fig. **17** is depicted as a two-dimensional diagram only for simplicity.

Plate thickness effects are incorporated into the analysis through the use of first order plate theory and the dislocation influence functions developed as detailed in the previous sections. It is therefore assumed that the stress components, crack opening displacement and plastic deformation are uniform across the thickness of the plate and are equal to the through-thickness average values. Further simplifications are made in the analysis with strain hardening and the Bauschinger effect, which have being neglected. This limits the models to situations where these factors do not greatly influence the crack opening loads and fatigue crack growth. Minor correction for these effects can be made; however, through the chosen value of the flow stress (e.g. use an average of the yield and ultimate strengths).

Steady State Fatigue Crack

First to be considered is the case of a fatigue crack that has been growing under constant amplitude loading for a significant number of cycles, such that the plastic wake is well established and the opening stress has stabilised [56].

In addition, it is assumed that any deviations in the loading history prior to the sustained period of constant amplitude loading have negligible effect on the current configuration. Following from these assumptions it can be seen that it is not necessary to model the entire crack growth all the way from initiation. At maximum applied load in the cycle, K_{max}, a zone of tensile plastic deformation, r_p, will be produced at the crack tip (x = 0) as depicted in Figs. **18a** and **18b**. Upon unloading towards minimum load, K_{min}, a compressive plastic zone, $r_{p,c}$, will result (Fig. **18c**). Propagation of the crack through the plastic zones will lead to the residual plastic material being left behind as an extra layer, or wake, along the crack faces (Figs. **18b** and **18c**). The load required to re-open the crack tip is referred to as the crack opening load, K_{op}. In these diagrams δ_M is the crack tip stretch at maximum load, δ_w is the wake height and r_{cc} is the edge of the region of crack closure.

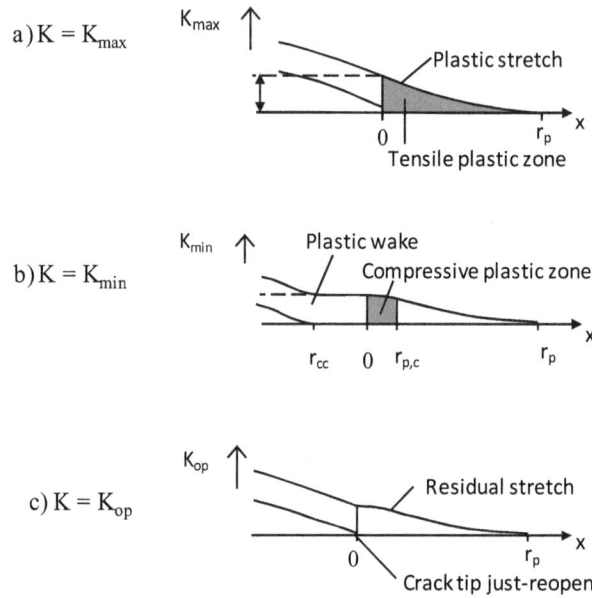

Figure 18: Schematic of steady state fatigue crack.

In a similar manner to as presented in the previous sections, the cracked region and zones of plastic deformation can be modelled by a continuous distribution of edge dislocations. The dislocations are chosen such that the resultant stress field is the same as the original crack problem. Along the crack length, the displacement function, $\delta(\xi)$, refers to the sum of the crack opening displacement and the plastic wake. Ahead of the crack tip, $\delta(\xi)$ is the plastic stretch curve. In both cases, $\delta(\xi)$ is related to the dislocation density function, $B_y(\xi)$, through equation (12), which is::

$$B_y(\xi) = -\frac{d\delta(\xi)}{d\xi} \tag{51}$$

When a mode I stress intensity factor, K, is remotely applied the resultant stress field is:

$$\frac{N_{yy}(x)}{2h} = \frac{1}{\pi}\int_{-\infty}^{r_p} B_y(\xi)G_{yy}(x;\xi)d\xi \tag{52}$$

The integral (52) is now separated into two parts, one over the crack length and the other over the tensile plastic zone. This leads to the equation:

$$\frac{N_{yy}(x)}{2h} = \frac{1}{\pi}\int_{-1}^{1} B_y(s)G_{yy}(x;s)\frac{2}{(1+s)^2}ds + \frac{1}{\pi}\int_{0}^{r_p} B_y(\xi)G_{yy}(x;\xi)d\xi \tag{53}$$

where along $-\infty < \xi \leq 0$ the following transformation has been applied:

$$\xi = \frac{s-1}{s+1}.$$

(54)

The first integral term $(-1 \leq s \leq 1)$ in (53) will be solved via the application of Gauss-Chebyshev quadrature (GC), while the second $(0 \leq \xi \leq r_p)$ will be determined through direct placement of the edge dislocations. This provides the stress function as a summation series:

$$\frac{N_{yy}(x)}{2h} = \frac{2}{m}\sum_{i=1}^{m}\phi(s_i)G_{yy}(x;s_i)\frac{1}{1+s_i} + \frac{1}{\pi}\sum_{j=1}^{n}\beta_y(\xi_j)G_{yy}(x;\xi_j)$$

(55)

where $\beta_y(\xi)$ is the Burger's vector of a dislocation located at ξ and is defined through use of Eq. (51) as $\beta_y(\xi) = -\Delta\delta(\xi) = \delta(\xi) - \delta(\xi + \Delta\xi)$ for $\Delta\xi > 0$. The non-singular function, $\phi(s)$, is part of the Gauss-Chebyshev procedure and is related to the dislocation density function through equation (37), which as a reminder is defined by:

$$B_y(s) = \phi(s)(1+s)^{1/2}(1-s)^{-1/2}.$$

(56)

Use has been made of the fact that as $s \to -1$ (or $\xi \to -\infty$) the dislocation density will approach zero. Therefore bounded at $s = -1$ quadrature is utilised. At $s = 1$ (or $\xi = 0$), however, the approximation of singular quadrature is made in order to keep the final system of linear equations determinate [17].

The reason for applying direct summation in crack tip plastic zone, rather then applying GC quadrature, is because for the latter approach the location of the integration points are pre-determined as a sinusoidal function. Consequently, the possible locations for the edges of the various zones, for example the compressive yield zone, will be limited to these point locations. This has negligible effect for the regions along the actual crack length, but can significantly affect the results for the deformation ahead of the crack tip. This will become of particular importance in the subsequent section when variable loading is considered. Furthermore, by placing the dislocations directly along the section $0 \leq \xi \leq r_p$, the locations of the points, ξ_i, and the mesh density can be chosen as desired. A similar approach was utilised by Nowell [57] who employed dislocation boundary elements over the entire length of the crack under large scale yielding conditions.

In (55), the GC integration points are defined in a similar manner to (21) as:

$$s_i = \cos\left[\pi\frac{2i-1}{2m}\right], \qquad i = 1...m$$

(57)

and the direct dislocation integration points by:

$$\xi_j = r_p\left[1 - \frac{j}{n+1}\right], \qquad j = 1...n,$$

(58)

Although for the latter, any chosen function can be used. When $x > r_p$ or $y \neq 0$, the stress function (54) may be evaluated at any point. However, within the crack and plastic zone interval, where $x < r_p$ and $y = 0$, (54) is only valid at the collocation points. For the GC quadrature these are:

$$t_k = \cos\left[\pi\frac{k}{m}\right], \qquad k = 1...m-1,$$

(59)

with

$$x = \frac{t-1}{t+1}.$$ (60)

For the direct quadrature, $0 < x < r_p$, the collocation points are taken as the halfway point between consecutive integration points, or mathematically:

$$x_j = \frac{\xi_{j+1} + \xi_j}{2}$$ (61)

In the case of the uniform distribution given by (57) this provides:

$$x_j = r_p\left[1 - \frac{j+0.5}{n+1}\right], \qquad j = 1... \, n.$$ (62)

The unknown functions ϕ and β_y are found by enforcing the known boundary conditions along the various sections of the crack interval. Within the plastic zone the material will yield in either tension or compression, or remain unchanged. Using a Tresca yield criterion this gives the boundary conditions for $0 < x, \xi < r_p$ as:

$$\beta_y(\xi_j) = \text{no change, no yielding},$$ (63a)

$$\frac{N_{yy}(x_j,0)}{2h} - \frac{N_{zz}(x_j,0)}{2h} = \sigma_f, \text{ tensile yielding},$$ (63b)

$$\frac{N_{yy}(x_j,0)}{2h} = -\sigma_f, \text{ compressive yielding}.$$ (63c)

It has been assumed that the out-of-plane constraint is minimal during compressive yielding and thus yielding occurs when $|N_{yy}/2h| = \sigma_f$. This approximation has been made in many past efforts to model crack closure using the strip-yield model (see for example Newman [51] and Wang and Blom [54]). For $x, \xi < 0$, the crack faces will either be open and traction free or closed, which provides the boundary conditions as:

$$\frac{N_{yy}(x_j,0)}{2h} = 0, \text{ crack open},$$ (64a)

$$B_y(\xi_i) = 0, \text{ crack closed}$$. (64b)

The known boundary conditions are applied to (55) within the appropriate regions along the crack length and plastic zone as depicted in Fig. **18**. This will provide a system of m - 1 + n linear equations in m + n unknowns. The last equation needed to complete the system comes from the side condition of a remotely applied stress intensity factor. By using the distributed dislocation technique the remote stress intensity factor can be applied directly through the displacement gradient, $d\delta(\xi)/d\xi$, without any prior knowledge of the actual displacement, as would be the case with boundary elements. The remote stress intensity factor is applied through the use of (26) as:

$$\phi(s_m) = \frac{K_I}{\sqrt{2\pi}} \frac{4}{E}$$ (65)

The resulting system of equations can be solved via any standard methods. An initial guess is made for the sizes of each of the various zones along the crack, which are r_p, $r_{p,c}$ and r_{cc} (see Fig. **18**). The calculated solution is then checked to ensure that all necessary conditions are satisfied in accordance with the applied boundary conditions (63) and (64). That is, the stresses in the plastic zone must be within the ranges $N_{yy} - N_{zz} \leq 2h\sigma_f$ and $N_{yy} \geq -2h\sigma_f$, and along the actual crack length $N_{yy} \leq 0$ for the crack to be closed. An iterative procedure is then employed until the final solution is reached; see Codrington and Kotousov [38], [56] and [58] for the full details.

The final and most important load case is the opening load, K_{op}, which is taken as the point when the crack tip just starts to re-open after minimum load. Boundary conditions for this case can be derived from the previous conditions (63) and (64) as:

$$\beta_y(\xi_j) = \beta_y^{max}(\xi_j), \quad 0 < \xi_j < r_p, \tag{66a}$$

$$N_{yy}(x_k, 0) = 0, \quad x_k < 0, \tag{66b}$$

where the superscript max refers to the results from the maximum load configuration.

Fatigue Crack Subjected to Variable Loading Sequence

The steady state analysis provides valuable insight into the effects of plasticity-induced crack closure on constant amplitude loading. However, most engineering structures will experience some form of variable loading, such as overloads and underloads. These divergences from the baseline load level will lead to stages of crack growth retardation or acceleration, depending on the load pattern that is applied. In this section, we now consider the case of variable loading following on from a sustained period of constant amplitude loading. The starting point for the present analysis will therefore be based on the results from the previous section for a steady state fatigue crack.

To investigate the effects of variable amplitude loading it is necessary to incrementally grow the crack and then determine the crack displacement and plastic deformation at each new length. For the sake of computational efficiency, the crack closure calculations are not carried out for every single load cycle. Instead, the crack is extended a given increment, Δa, over which the crack opening load is held constant. The number of load cycles needed to extend the crack is found by stepping through the load sequence cycle-by-cycle and summing up the growth for each individual cycle. Any typical crack growth law may be utilised by rewriting it in terms of the effective stress intensity factor range, such that:

$$\frac{da}{dN} = f(\Delta K_{eff}). \tag{67}$$

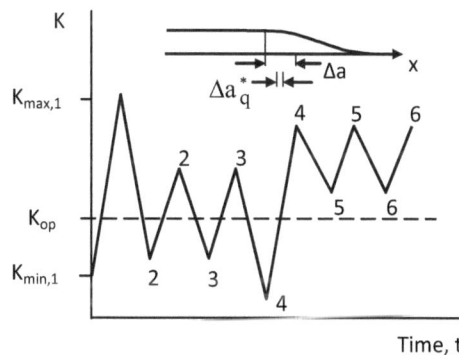

Figure 19: Example variable load sequence.

For a single load cycle, the previous equation (67) is equivalent to the crack growth increment for one cycle, Δa_q^*, where q refers to the cycle number post steady state analysis (see Fig. **19**). The effective stress intensity factor range is then defined by:

$$\Delta K_{eff,q} = \begin{cases} K_{max,q} - K_{op}, & \text{if } K_{min,q} < K_{op} \tag{68a} \\\\ K_{max,q} - K_{min,q}, & \text{if } K_{min,q} > K_{op} \tag{68b} \end{cases}$$

When the sum of the single cycle growth increments, $\Sigma \Delta a_q^*$, is greater than or equal to the crack increment, Δa, the closure model is implemented.

The procedure for the variable analysis starts by taking the results for the opening load as found from the steady state analysis. This value is used to determine the block of load cycles for the first crack increment. During the crack growth calculations the load sequence is monitored to find the highest maximum load, $K_{max,H}$, and the lowest minimum loads applied before, $K_{min,B}$, and after, $K_{min,A}$, the highest load. For the example cycle sequence given in Fig. **17** this would provide $K_{max,H} = K_{max,1}$, $K_{min,B} = K_{min,1}$, and $K_{min,A} = K_{min,4}$. These values are then used to determine the crack opening displacement and plastic stretch curves, and hence the crack opening load. Subsequent values for the crack opening load stress intensity factor are found by implementing the distributed dislocation model. In the current work, the crack closure analysis was executed based on the following algorithm

1. Apply $K_{min,B}$ at current crack length (provided $K_{min,B} <$ previous $K_{min,A}$),

2. Apply $K_{max,H}$ at current crack length,

3. Extend crack by an increment of Δa,

4. Apply lowest minimum load, $K_{min,A}$, at new crack length,

5. Determine crack opening load, K_{op}, at new crack length,

6. Calculate the block of load cycles for the next crack increment,

7. Repeat the process.

If an overload cycle, or highest maximum cycle, is applied in the first load block, the initial analysis follows the same procedure as for maximum load in the previous steady state crack section. This will provide a new tensile plastic zone, say $r_{p,ov}$, and associated dislocation mesh ahead of the crack tip (at $x = r_{tip} = 0$). As the crack continues to propagate under variable amplitude loading, the stress field and the resulting distribution of plastic deformation, will become quite complex. Therefore it is most ideal to apply the boundary conditions in this region on an individual dislocation basis, rather then by a 'zone' basis as used for the steady state analysis. The stress conditions are applied to the collocation points, x_j, through equation (55) and the displacement boundary conditions are applied to the integrations points, ξ_j, and will now be applied in terms of the actual displacement/plastic stretch via:

$$\delta(\xi_d) = \sum_{j=1}^{d} \beta_y(\xi_j), \quad 0 < \xi < r_p, d = 1 \text{ n}. \tag{69}$$

The solution procedure for the variable loading model follows by enforcing the known boundary conditions over the length of the crack and plastic zones. For the plastic zone ahead of the crack tip, at $x = r_{tip}$ where $r_{tip} \geq 0$, there are three possibilities. The material may yield in either compression or tension, or the residual stretch may remain unchanged. In the case of no yielding, this leads to the conditions for $r_{tip} < x$, $\xi < r_p$, as:

$$\delta(\xi_j) = \delta^{prev}(\xi_j), \tag{70a}$$

$$-\sigma_f < \frac{N_{yy}(x_j,0)}{2h}, \tag{70b}$$

$$\frac{N_{yy}(x_j,0)}{2h} - \frac{N_{zz}(x_j,0)}{2h} < \sigma_f. \tag{70c}$$

When tensile yielding occurs, then:

$$\frac{N_{yy}(x_j,0)}{2h} - \frac{N_{zz}(x_j,0)}{2h} = \sigma_f, \tag{71a}$$

$$\delta(\xi_j) > \delta^{\text{prev}}(\xi_j) \, , \tag{71b}$$

and lastly, if there is compressive yielding:

$$\frac{N_{yy}(x_j, 0)}{2h} = -\sigma_f \, , \tag{72a}$$

$$\delta(\xi_j) < \delta^{\text{prev}}(\xi_j) \, . \tag{72b}$$

The superscript prev refers to the displacement calculated from the previous load configuration. Similar conditions apply to the incremented crack tip region, $0 < x, \xi < r_{\text{tip}}$, however now there is the possibility that the crack may be open, which provides the following conditions:

$$N_{yy}(x_j, 0) = 0 \, , \tag{73a}$$

$$\delta(\xi_j) > \delta^{\text{prev}}(\xi_j) \, , \tag{73b}$$

if crack is open:

$$\delta(\xi_j) = \delta^{\text{prev}}(\xi_j) \, , \tag{74c}$$

$$-\sigma_f < \frac{N_{yy}(x_j, 0)}{2h} < 0 \, , \tag{74d}$$

for the case when crack is closed, and in the case of the compressive yielding:

$$\frac{N_{yy}(x_j, 0)}{2h} = -\sigma_f \, , \tag{75a}$$

$$\delta(\xi_j) < \delta^{\text{prev}}(\xi_j) \, , \tag{75b}$$

Along the region of the crack that was formed prior to the variable loading sequence (where $x, \xi < 0$), i.e. the region where the wake thickness is uniform; there are two possibilities. The crack may be open and traction free or the crack closed and in compression. In this section of the crack it is assumed that no yielding will occur. The boundary conditions for $x, \xi < 0$ therefore become:

$$N_{yy}(x_k, 0) = 0 \, , \tag{76}$$

for the crack to be open, and:

$$B_y(\xi_i) = 0 \, , \tag{77a}$$

$$\delta(\xi_i) = \delta_w \, , \tag{77b}$$

$$N_{yy}(x_k, 0) \le 0 \tag{77c}$$

if the crack is closed. The displacement can be found from the summation series:

$$\delta(\xi_e) = \frac{1}{2} \sum_{i=2}^{e} \left[B_y(\xi_j) + B_y(\xi_{j-1}) \right] \left(\xi_{j-1} - \xi_j \right) + \delta^*(\xi_n) \, , \tag{78}$$

where $e = 2...$ m, $\xi < 0$ and δ^* refers to the displacement distribution (69) at $d = n$.

For each load case an initial guess is made for the boundary conditions within each of the integration ranges. These are then assigned to the various collocation or integration points based on the methods described above to give a system of $m - 1 + n$ linear equations in the $m + n$ unknowns. The final equation again comes from the side condition of a remotely applied stress intensity factor, given by (26). After solving the system for a particular load case the results must be checked to determine whether or not any of the applied boundary conditions need changing.

As described in the previous section, the crack opening load is found by applying a load K_{op} and adjusting the value until the crack tip is on the verge of re-opening. The plastic deformation ahead of the crack tip for this load case will remain unchanged.

APPLICATIONS

In this section we describe some applications of the developed computational technique for the analysis of non-linear crack growth phenomena. First we investigate the fatigue crack growth under constant amplitude loading under small-scale yielding condition. The latter allows avoiding effects of the specimen's geometry and loading conditions, and as such, can be applicable to many practical situations. Further, we investigate the effect of a single overload cycle on the fatigue crack growth under constant amplitude. Finally, we consider the growth of small cracks from deep notches. An interesting result is that the fatigue crack growth rates are bounded by predictions based on plane stress and plane strain theories. A comparison of theoretical predictions for all three models with experimental studies demonstrates a good agreement indicating a great potential of the developed theory.

Constant Amplitude Fatigue Crack Growth

The traditional linear elastic approach to fatigue crack growth, using two-parameter crack growth rate relations, tends to result in large scatter amongst the growth rate data obtained from various studies, and for different load ratios and specimen thicknesses.

A parameter frequently utilised to describe the effects of the load ratio, R, on fatigue crack growth is the effective stress intensity factor range ratio, U. This parameter is defined as:

$$U = \frac{\Delta K_{eff}}{\Delta K} = \frac{1 - K_{op}/K_{max}}{1 - R},$$ (79)

where ΔK_{eff} is the effective stress intensity factor range and ΔK ($= K_{max} - K_{min}$) is the traditional linear elastic stress intensity factor range. The stress intensity factors K_{max} and K_{min} are the maximum and minimum values, respectively, experienced for the constant amplitude cyclic loading. The load ratio is therefore given by $R = K_{min}/K_{max}$. In the case when K_{min} is less than zero, then K_{min} is set equal to zero in Eq. (79).

It is evident that the classical plane theories of elasticity, which are based on plane stress or plane strain assumptions, are not suitable for quantitative assessment of the thickness effect. In this section, first-order plate theory and the steady state fatigue crack model developed in the previous sections are utilised to present a generalised formula for the effective stress range ratio. Under the assumption of small-scale yielding and for an ideal material (homogeneous and isotropic) with a fixed material Poisson's ratio, ν; the effective stress intensity range ratio, U, can only be function of the non-dimensional parameters R and χ where:

$$\chi = \frac{K_{max}}{\sigma_f \sqrt{h}}$$ (80)

The material properties parameter, σ_f / E, was found to have no effect on the crack opening load ratio, however, it does alter the crack opening displacement and plastic zones.

Theoretical results obtained for the ratio of the opening to applied stress intensity factor are presented in Fig. **20** as a function of the R ratio and the parameter χ (for $\nu = 0.3$). The results for both plane stress and plane strain are also given and indicated in the figure. It can be seen that all results for the finite thickness analysis are between the two limiting cases of plane stress and plane strain. The plane stress curve therefore represents the upper limit and the plane strain the lower limit. An increase of the plate thickness corresponds to a transition of the stress state and crack tip parameters from plane stress conditions to a plane strain state. The results also show that the opening stress intensity factor ratio increases with R ratio and K_{max} and decreases with the increase of the plate thickness and flow stress.

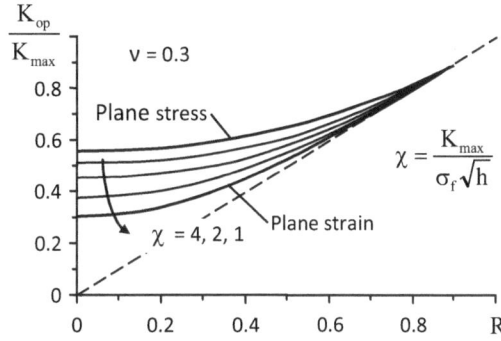

Figure 20: Ratio of the opening to applied stress intensity factors as a function of R ratio and parameter χ.

The plane stress solution for a steady-state propagating crack was first derived by Budiansky and Hutchison [49], based on the complex variable function approach. The theoretical curve obtained from their study is virtually indistinctive with the plane stress numerical results presented in Fig.20.

Using non-linear regression, the results for the effective stress intensity range ratio, U, under small-scale yielding conditions ($\nu = 0.3$), can be approximated by the following function

$$U(R,\chi) = A(\chi) + B(\chi)R + C(\chi)R^2 ,$$ **(81)**

with

$$A(\chi) = 0.446 + 0.266 \cdot e^{-0.41\chi} ,$$ **(82a)**

$$B(\chi) = 0.373 + 0.354 \cdot e^{-0.235\chi} ,$$ **(82b)**

$$C(\chi) = 0.2 - 0.667 \cdot e^{-0.515\chi} ,$$ **(82c)**

where parameter χ can also be written through the range of the stress intensity factor ΔK as

$$\chi = \frac{K_{max}}{\sigma_f \sqrt{h}} = \frac{\Delta K}{(1-R)\sigma_f \sqrt{h}}$$ **83)**

The approximation (81)-(82) is a fitting function to the theoretical results with the form of the coefficients A, B, and C chosen for simplicity and to give the correct behaviour at the plane stress and plane strain limits (i.e. as $\chi \to \infty$ and $\chi \to 0$, respectively). It was found that the fitting function (81) with the coefficients given by Eqs (82) produces an error of less than 5 % with the theoretical results when R < 0.8 for arbitrary plate thicknesses. The effective stress intensity range ratio determined from the regression fit (81) can be used to reduce the scatter in the constant amplitude fatigue crack growth data for different load ratios and plate thicknesses. An example of this is shown in Fig. **21** using the experimental data of Park and Lee [59] who tested type 304 stainless steel compact tension specimens.

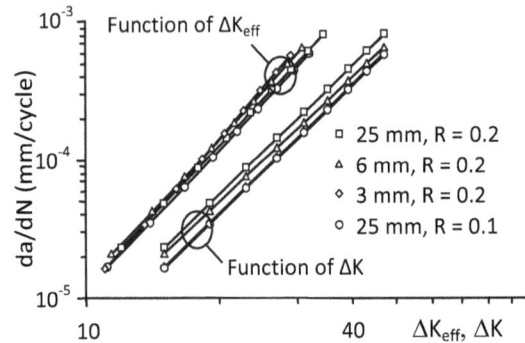

Figuee 21: Crack growth rate versus the stress intensity factor range (Park and Lee [59]) and the effective stress intensity factor range, ΔK_{eff} (present approach).

Three different plate thicknesses were tested (2h = 3, 6 and 25 mm) along with two load ratios (R = 0.1 and 0.2). In this figure, the crack growth rate data is plotted against the traditional linear elastic stress intensity factor range, ΔK, as well as the effective stress intensity factor range, ΔK_{eff}, as determined from the developed models. It can be seen that when plotted against the ΔK_{eff} the scatter in the experimental results is significantly reduced, without the need for any curve fitting of the experimental data.

Fatigue Crack Growth Retardation Following an Overload Cycle

Numerous past experiments (e.g. Elber [29]; Shin and Hsu [60], Shuter and Geary [43], Skorupa [31], Borrego et al. [44]) have shown that the application of a tensile overload cycle can provide substantial crack growth retardation and even complete crack arrest. This suggests the possible use of single or periodic overloads as a way of extending the fatigue life of a cracked component in service. There are, however, several factors that influence the post-overload fatigue behaviour including the load amplitude, material properties, crack geometry and plate thickness. The extent of the overload retardation is therefore dependent not only on the overload cycle and baseline loading, but also on the various factors listed above and the interaction between these parameters. In general, it has been shown that an increase in the plate thickness means a reduction in the extent of overload retardation (Shuter and Geary [43], Bichler and Pippan [61]).

The key feature of post-overload crack growth observed in most experimental studies is that of a delayed retardation effect. Immediately after the overload cycle there is a small period of accelerated growth, which is then followed by a reduction in the crack growth rate below the pre-overload value and down to a minimum. As the crack propagates further the growth rate steadily increases back to the pre-overload or constant amplitude growth rate appropriate to the new crack length. Some of the proposed mechanisms for the overload retardation are: residual stresses (Schijve and Broek [62]); crack closure (such as plasticity and roughness-induced) (Elber [29], Suresh [63], Shin and Hsu [60]); and crack branching or deflection (Suresh [64]).

For pre-overload growth rates that are well above threshold, it is commonly found that plasticity-induced crack closure is the primary mechanism of overload retardation (Elber [29], Shin and Hsu [60], Shuter and Geary [43], Borrego et al. [43], Bichler and Pippan [61]). The application of an overload cycle will result in a larger tensile plastic zone ahead of the crack, when compared to the pre-overload zone, as well as plastic blunting of the crack tip. The overload tensile plastic deformation will also cause an increase in the size of the compressive plasticity zone at the crack tip once the load is removed. With continued crack growth the overload plastic deformation will remain as a 'lump' in the plastic wake left on the crack faces. These contributions will alter the load required to fully open the crack and thus affect the crack growth rate.

The theoretical framework of the previous section will now be utilised to investigate the situation of a single tensile overload in otherwise constant ΔK loading. This idealisation is of great practical importance as overloads are common place in many real structures, e.g. in an aircraft during take-off and landing. The effect of the overload ratio on the crack opening load is displayed in Figs. **22** and **23** for the case of a plane stress analysis. Here the overload stress intensity factor is K_{ov}, the overload ratio is K_{ov}/K_{max}, and K_{max} and K_{min} are the maximum and minimum values, respectively, of the constant ΔK loading. The crack extension in Fig. **22** is normalised by the plane stress tensile overload plastic zone size, $r_{p,ov}$. It can be seen that there is an initial drop in the opening load immediately following the overload cycle, which corresponds to an increase in the crack growth rate. The crack opening load then steadily increases up to a peak value before decreasing back down and tending to the steady state value. The section of the curve where the opening load ratio is greater than that of the steady state value is what causes the retardation in the crack growth. Fig. **23** shows the peak opening load ratio as a function of the overload ratio and the R ratio. Lines of constant $\Delta K_{ov}/\Delta K$ are also indicated as this is an alternative definition for the overload ratio. With an increase in the overload ratio, there is an accompanying increase in the peak opening load ratio and the distance to recover the steady state value. The peak value provides a good indication of the overall amount of growth retardation and demonstrates the potential for complete crack arrest (i.e. when ΔK_{eff} is less than or equal to the long crack threshold, $\Delta K_{eff,TH}$).

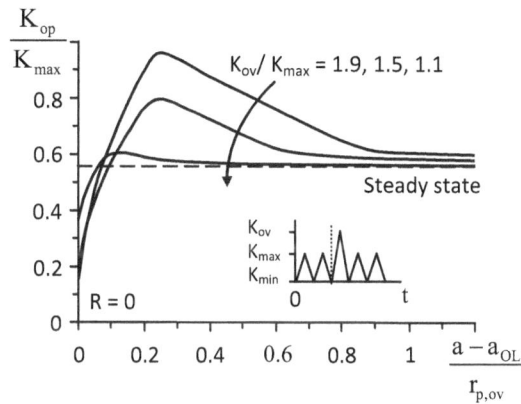

Figure 22: Plane stress crack opening load ratio following the application of an overload cycle as a function of crack extension.

Figs. **24** and **25** display the results for the crack opening load ratio as determined using the finite thickness model. In Fig. **24**, the crack extension is normalised by the finite thickness overload plastic zone size, which is calculated using the theoretical methods described previously and for $v = 0.3$ can be approximated by the curve fit:

$$r_{p,ov} = \frac{K_{ov}^2}{\sigma_f^2}\left(-0.124\exp(-0.552\chi^{1.95}) + 0.192\exp(-4.16\chi^{-1.99}) + 0.199\right). \tag{84}$$

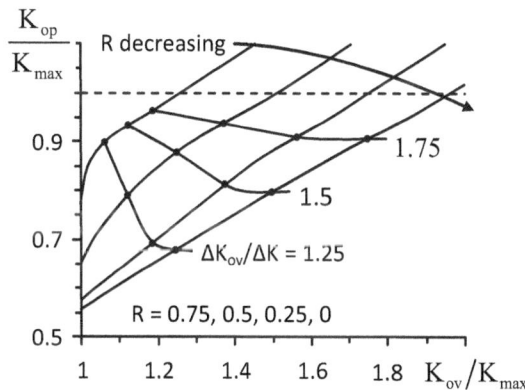

Figure 23: Plane stress crack opening load ratio following the application of an overload cycle as a function of peak value.

Curves for several values of the non-dimensional parameter χ are provided. It can be seen in both figures that a decrease in the parameter χ leads to a reduction in the change in opening load ratio. Therefore, for a fixed K_{max}, R and σ_f an increased plate thickness will result in less crack growth retardation.

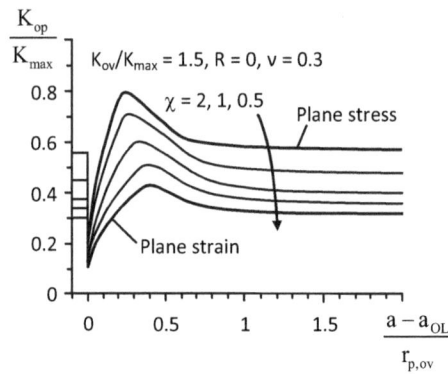

Figure 24: Thickness effect on the crack opening load ratio following the application of an overload cycle as a function of crack extension.

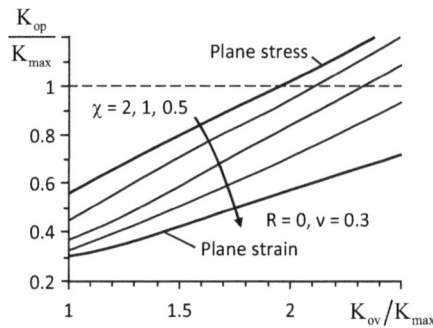

Figure 25: Thickness effect on the crack opening load ratio following the application of an overload cycle as a function of peak value.

The crack opening load ratio can be used to predict the fatigue crack growth following the application of an overload cycle. To demonstrate the adequacy of the developed approach, the theoretical predictions will be compared with previously published experimental data. The first experimental results to be considered are those of Borrego *et al.* [44] who undertook single overload tests on 6082-T6 aluminum alloys.

Center cracked tension specimens of 3 mm thickness were loaded under constant ΔK loading before and after the overload cycle. The yield strength of the material is 245 MPa and ultimate strength is 300 MPa. The Paris growth law is utilised with the 'effective' material constants found by fitting data from Borrego *et al.* [65] using values for the steady-state crack opening load ratio determined from the present theoretical methods. These constants are $C_{eff} = 1.05 \times 10^{-7}$ and $m = 3.92$ for ΔK_{eff} in MPa m$^{1/2}$ and da/dN in mm/cycle.

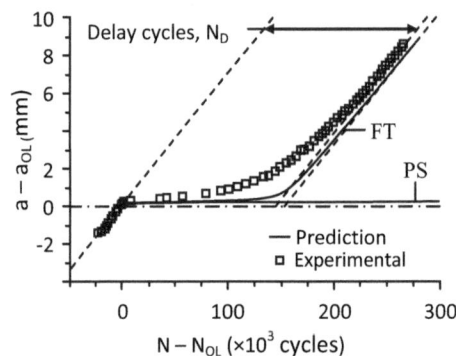

Figure 26: Comparison of the plane stress (PS) and finite thickness (FT) predictions for the crack growth after an overload cycle.

Figs. **26** and **27** show a comparison between predictions made with plane stress and finite thickness dislocation influence functions. In these tests $\Delta K = 8$ MPa m$^{1/2}$, $R = 0.05$, $K_{ov}/K_{max} = 1.95$, and in the finite thickness case $\nu = 0.3$ and $2h = 3$ mm.

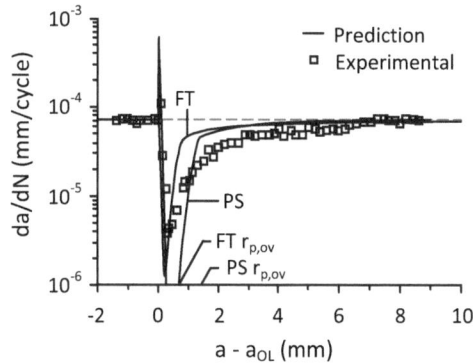

Figure 27: Comparison of the plane stress (PS) and finite thickness (FT) predictions.

The plane stress prediction severely overestimates the retardation effect, while the finite thickness prediction provides a very good estimate. In both cases, however, the predicted growth rate returns towards the pre-overload rate within the respective overload plastic zones. The experimental data shows continued retardation well outside of the plastic zone. This can be partly attributed to strain hardening (Pommier and de Freitas [66]), which is neglected in the present analysis. Further results illustrating the effect of the baseline ΔK are given in Figs. **28** and **29** using the finite thickness analysis. A good correlation can be seen between the predictions and experimental values. One measure of the extent of the retardation effect is the number of delay cycles produced by the overload cycle. The number of delay cycles refers to the number of load cycles required to restore the pre-overload crack growth rate minus the number of cycles needed to reach the same crack length without an overload applied (see Fig. **26**).

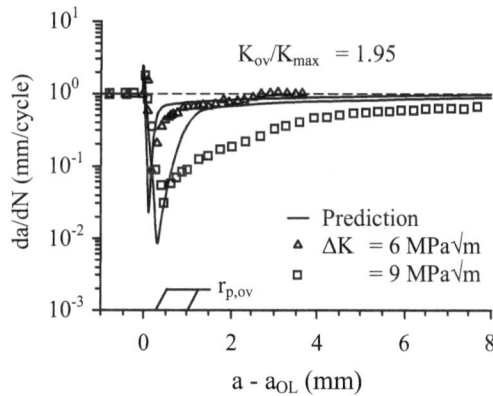

Figure 28: Effect of ΔK ratio on crack growth rate versus crack length post-overload.

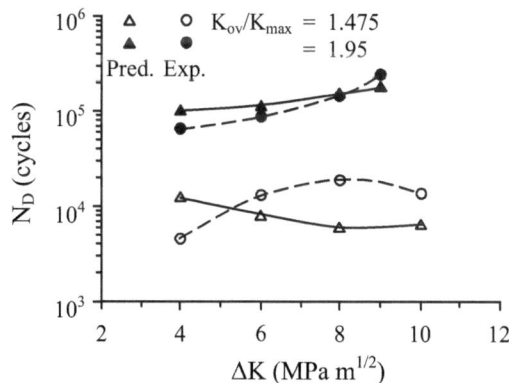

Figure 29: Effect of plate thickness on crack growth rate versus crack length post-overload.

The results for the number of delay cycles in Fig. **29** show that overall, the developed models can be used to provide an adequate prediction for the effect of overload ratio and baseline ΔK.

The effect of specimen thickness on the crack growth retardation after a tensile overload cycle was investigated by Shuter and Geary [43]. They used compact tension specimens of various thicknesses machined from BS 4360 Grade 50D carbon-manganese structural steel. The yield strength of this material is 396 MPa and the ultimate strength is 545 MPa. Tests were conducted at a baseline $\Delta K = 25$ MPa m$^{1/2}$ with R = 0.1 and $K_{ov}/K_{max} = 2$. A comparison of the experimental results with the predicted crack growth rate curves is given in Fig. **30** for specimen thicknesses of 2h = 7.5 mm and 25 mm. Fig. **31** provides the obtained number of delay cycles as a function of specimen thickness. In these calculations the effective Paris law constants are $C_{eff} = 6.0 \times 10^{-8}$ and m = 2.7, which have been adapted from general data available in the literature for this material (e.g. Thorpe *et al.* [67]).

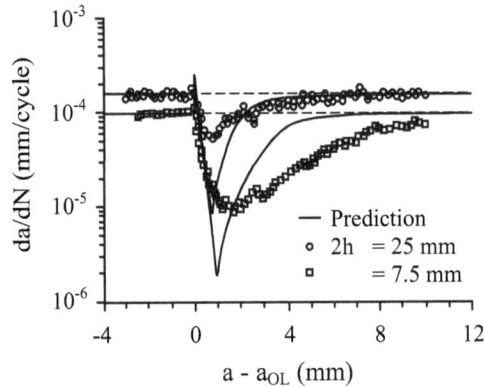

Figure 30: Effect of plate thickness on the number of delay cycles.

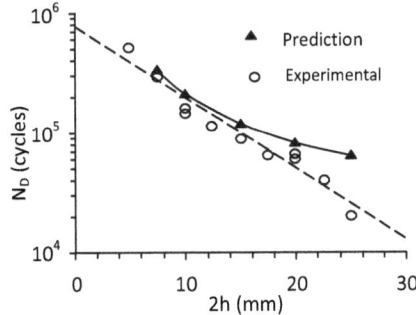

Figure 31: Effect of plate thickness on the number of delay cycles.

The predicted and experimental growth rate curves display the same trends with an increase in specimen thickness leading to a decrease in the retardation. A reasonable agreement is also seen with the predicted number of delay cycles, particularly for the thinner specimens, though the predictions are consistently greater than the measured values.

Short Fatigue Crack Growth from Sharp Notches

The initiation of fatigue cracks in structural components usually occurs at stress concentrators like holes or notches. Such geometrical features may be part of the component design (bolt holes, key ways, weld toes, etc) or could be defects introduced during manufacturing (welding defects, surface scratches, etc). These stress concentrators have a significant influence on the early stages of fatigue crack growth and therefore need to be incorporated into predictions made for the service-life of a component. Past experimental studies (Leis [68], Shin and Smith [69], Benedetti *et al.* [70]) have found that small cracks emanating from holes and notches tend to propagate at much faster rates in comparison to long established cracks under the same applied loading. In addition, small cracks can grow at an applied elastic stress intensity factor range, ΔK, well below that of the long crack threshold, ΔK_{th}.

Research into the behaviour of small cracks has provided a range of possible explanations and governing mechanisms. During the initial stages of growth, when the crack size is comparable to the microstructure, factors such as the local grain orientation, small particles and inclusions will have a great influence (Taylor and Knott [71], Suresh and Ritchie [72]). As the crack increases in size, however, these effects will become less significant and the growth rate will be an average over several grain lengths. For cracks of a length greater than any significant microstructural features, plasticity-induced crack closure has been found able to explain the various stages of short crack growth (Liaw and Logsdon [73]). A newly initiated crack will have a limited growth history and thus a significant plastic wake will not yet be developed. Small cracks will therefore grow at faster rates than long cracks where the plastic wake is fully developed.

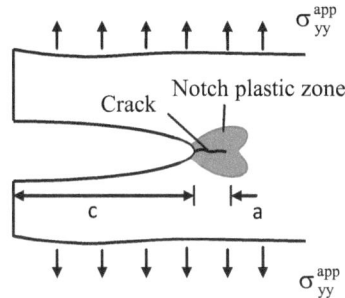

Figure 32: Schematic of small fatigue crack growth from notch.

The methods of the previous sections will now be extended to consider the fatigue crack growth from sharp notches. In the case of a small crack emanating from a long and narrow notch in a large plate, the notch geometry can be well approximated by an edge crack of the same length (as depicted in Fig. **32**).

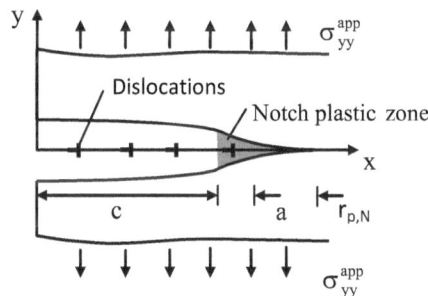

Figure 33: Simplified dislocation representation of the problem.

Here, the notch depth is given by c, the crack length by a, $r_{p,N}$ is the length of the original notch plastic zone, and σ_{yy}^{app} is the remotely applied cyclic stresses. No initial plastic wake is assumed to exist along the notch length as this would contradict the small crack hypothesis. By replacing the notch and crack geometry with edge dislocations (see Fig. **33**), the resultant stress along the y-axis can be obtained via:

$$\sigma_{yy}(x) = \frac{E}{4\pi D} \int_0^A b_y(\xi) K(x;\xi)\, d\xi + \sigma_{yy}^{app}(x),$$

(85)

where D = 1 for plane stress and $D = 1 - v^2$ for plane strain, and σ_{yy}^{app} is the remotely applied stresses. For the case of an edge crack, the plane stress/strain influence function K(x; ξ) is given by [17]:

$$K(x;\xi) = \frac{1}{x - \xi} - \frac{1}{x + \xi} - \frac{2\xi}{(x + \xi)^2} + \frac{4\xi^2}{(x + \xi)^3}$$

(86)

Initially, the 'equivalent' crack length, A, in equation (85) is given by the notch length plus notch plastic zone, that is $A = c + r_{p,N}$. As the crack propagates the equivalent crack length will become the sum of the notch length, small crack length and crack tip plastic zone, r_p (i.e. $A = c + a + r_p$). The solution procedure then follows as outlined in the previous sections.

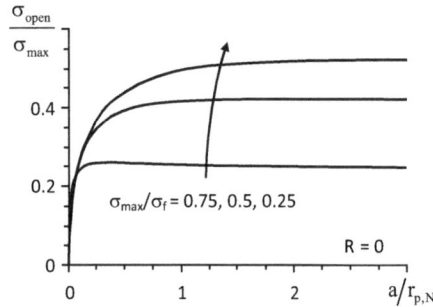

Figure 34: Variation in the plane stress crack opening stress ratio as a function of the maximum applied stress to flow stress ratio.

Results for the variation in the opening stress ratio, $\sigma_{open}/\sigma_{max}$, as the crack propagates are shown in Figs. **34** and **35**. These values are for plane stress conditions and are given as a function of the maximum applied stress to flow stress ratio, σ_{max}/σ_f, (Fig. **34**) and as a function of the load ratio, R (Fig. **35**). The crack length, a, has been normalised by the initial notch plastic zone size, $r_{p,N}$. It can be seen that as the crack length increases the crack opening stress also increases towards a steady state value. This behaviour is as expected since there will be a gradual build up of residual plastic material on the newly formed crack faces.

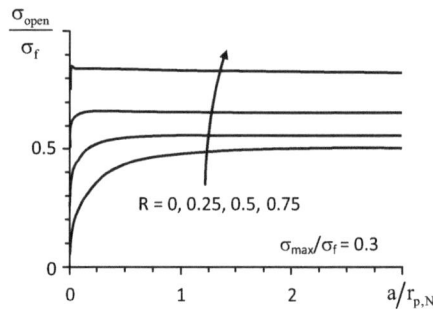

Figure 35: Variation in the plane stress crack opening stress ratio as a function of the load ratio.

The results for the transient behaviour under plane strain conditions show the same trends as for the plane stress state. Thus detailed results are omitted. A summary of the steady state values for plane stress and plane strain, and hence the bounds for the finite thickness case, are provided in Figs. **36** and **37**.

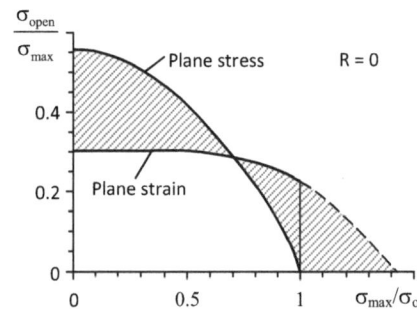

Figure 36: Bounds for the steady state crack opening stress ratio as a function of the maximum applied load to flow stress ratio.

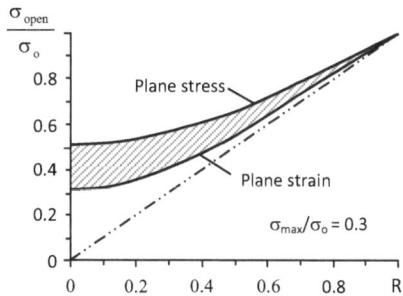

Figure 37: Bounds for the steady state crack opening stress ratio as a function of the load ratio.

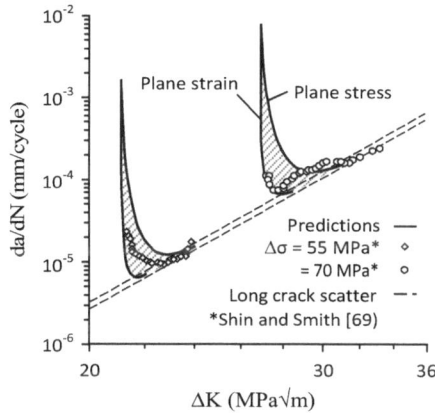

Figure 38: Prediction bounds for the fatigue crack growth rate of small cracks in stainless steel as a function of the stress intensity factor range.

These results show that for low to medium values of σ_{max}/σ_o fatigue cracks will grow faster in thicker specimens where the stress state is nearer to plane strain. Fig. **36** implies that there is a reversal of this behaviour at $\sigma_{max}/\sigma_o \sim 0.7$, although this requires further investigation.

The experimental results of Shin and Smith [69] for the fatigue crack growth from sharp notches will now be considered. Fatigue tests where undertaken on double edge notched AISI 316 stainless steel specimens with a thickness of 2.6 mm. Elliptical notches of a length of 35 mm were spark-machined into the specimens for several notch root radii.

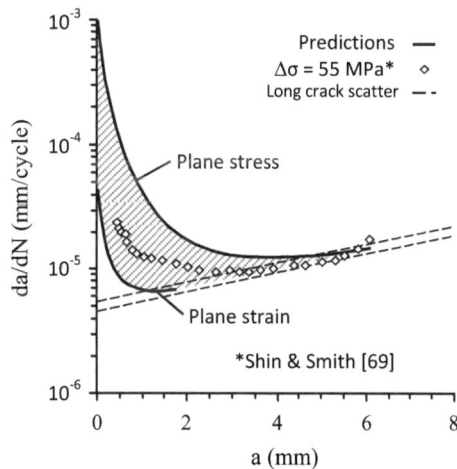

Figure 39: Prediction bounds for the fatigue crack growth rate of small cracks in stainless steel as a function of the crack length.

The data considered here are for the notch root radius of 0.4 mm. The material properties for the stainless steel are a yield stress of 256 MPa and an ultimate strength of 574 MPa. Tests were conducted under constant amplitude loading with a load ratio of R = 0.05. Results for the plane stress and plane strain predictions of the crack growth rate for two load cases are given in Figs. **38** and **39** along with the experimental values. The hatched area in these figures indicates the prediction range between the plane stress and plane strain bounds. It can be seen that the plane stress prediction provides an upper bound while the plane strain provides a lower bound for both of these cases. Initially the experimental values are nearer the plane strain predictions, but as the crack length increases they tend towards the plane stress values.

REFERENCES

[1]   Hartranft RJ, Sih GC. An approximate three-dimensional theory of plates with application to crack problems. Int J Eng Sci 1970; 8: 711-729.

[2]   Kwon SW, Sun CT. Characteristics of three-dimensional stress fields in plates with a through-the-thickness crack. Int J Fracture 2000; 104: 291-315.

[3]   Kotousov A. Fracture in plates of finite thickness. Int J Solids Structures 2007; 44: 8259-8273.

[4]   Kotousov AG, Harding S, Lazzarin P. On plate thickness effect in plane problems of elasticity. In proc: Crack Path 2009, Vicenza, Italy.

[5]   Kane TR, Mindlin RD. High frequency extensional vibrations of plates. J App Mechanics 1956; 23: 277-283.

[6]   Kotousov A, Wang CH. Three-dimensional stress constraint in an elastic plate with a notch. Int J Solids Structures 2002; 39: 4311-4326.

[7]   Timoshenko S, Woinowsky-Krieger S. Theory of plates and shells, Second Edition, 1959, McGraw Hill.

[8]   Yu Y. Vibration of elastic plates: linear and non-linear dynamical modelling of sandwiches, laminated composites, and piezoelectric layers, Springer-Verlag, New York, 1995.

[9]   Yang W, Freund LB. Transverse shear effects for through cracks in an elastic plate. Int J Solids Struct 1985; 21: 977-994.

[10]   Timoshenko SP, Goodier JN. Theory of elasticity, Third Edition, 1970, McGraw-Hill.

[11]   Kotousov A, Wang CH. Fundamental solutions for the generalized plane strain theory. Int J Eng Sci 2002; 40:1775-1790.

[12]   Liebfried G. Verteilung von Versetzungen im statischen Gleichgewicht. Zeitschrift fur Physik 1951; 130: 214-226.

[13]   Eshelby JD. The continuum theory of lattice defects. Solid State Physics 1956; 3: 79-144.

[14]   Erdogan F. Fracture Mechanics, Int J Solids Structures 2000; 37: 171-183.

[15]   Keer LM, Mura T. Stationary crack and continuous distribution of dislocations, in Proceedings of the 1st International Conference on Fracture, 1966; 99-115.

[16]   Codrington J, Kotousov A. Application of the distributed dislocation technique for calculating crack tip plasticity effects. Fatigue Fract Eng Mater Struct 2007; 30: 1182-1193.

[17]   Hills DA, Kelly PA, Dai DN, Korsunsky AM. Solution of crack problems. The distributed dislocation technique. Kluwer Academic Publisher, The Netherlands, 1996.

[18]   Erdogan F, Gupta GD. On the numerical solution of singularintegral equations. Q. Appl. Math. 1972; 30: 525-534.

[19]   Nakamura T, Parks DM. Three-dimensional stress field near the crack front of a thin elastic plate. J Appl Mech 1988; 55: 805-813.

[20]   She C, Guo W. The out-of-plane constraint of mixed-mode cracks in thin elastic plates. Int J Solids Struct 2007; 44: 3021-3034.

[21]   Pfaff RD, Washabaugh PD, Knauss WG. An interpretation of Twyman-Green interferograms from static and dynamic fracture experiments. Int J Solids Struct 1994; 32: 939–955.

[22]   Rice JR, A path independent integral and the approximate analysis of strain concentration by notches and cracks. J Appl Mech 1968; 35: 379–386.

[23]   Humbert L, Valle V, Cottron M. Experimental determination and empirical representation of out-of-plane displacements in a cracked elastic plate loaded in mode I. Int J Solids Struct 2000; 37: 5493-5504.

[24]   Dugdale DS. Yielding of steel sheets containing slits. J Mech Phys Solids 1960; 8: 100–104.

[25]   Kotousov A, Wang CH. Strip yield model for a crack in a plate of arbitrary thickness, The 14th European Conference on Fracture (ECF14), 8-13 September 2002, Krakow, Poland: 217-224.

[26]   Kelly PA, Nowell D. Three-dimensional cracks with Dugdale-type plastic zones. Int J Fract 2000; 106: 291–309.

[27]   Newman Jr JC, Crews Jr JH, Bigelow CA, Dawicke DS. Variations of a global constraint factor   in cracked bodies under tension and bending loads. Constraint effects in fracture theory and applications, ASTM STP 1995; 1244: 21–42.

[28]   Elber W. Fatigue crack closure under cyclic tension. Eng Fract Mech 1970; 2: 37-45.

[29]   Elber W. The significance of fatigue crack closure. Damage Tolerance in Aircraft Structures, ASTM STP 1971; 486: 230-242.

[30]   Newman Jr JC. The merging of fatigue and fracture mechanics concepts: a historical perspective. Prog Aerosp Sci 1998; 34: 347-390.

[31]   Skorupa M. Load interaction effects during fatigue crack growth under variable amplitude loading – a literature review. Part II: qualitative interpretation. Fatigue Fract Eng Mater Struct 1999; 22: 905-926.

[32]   Dougherty JD, Srivatsan TS, Padovan J. Fatigue crack propagation and closure behavior of modified 1070 steel: experimental results. Eng Fract Mech 1997; 56: 67–187.

[33]   Skorupa M, Skorupa A. Experimental results and predictions on fatigue crack growth in structural steel. Int J Fatigue 2005; 27: 1016-1028.

[34]   Schijve J. Some formulas for the crack opening stress level. Eng Fract Mech 1981; 14: 461-465.

[35]   McEvily AJ, Ishihara S. On the development of crack closure at high R levels after and overload. Fatigue Fract Eng Mater Struct 2002; 25; 993-998.

[36] Bachmann V, Munz D. Crack closure in fatigue of titanium alloy. Int J Fract 1975; 11: 713-716.

[37]   James MN, Pacey MN, Wei LW, Patterson EA. Characterisation of plasticityinduced closure-crack flank contact force versus plastic enclave. Eng Fract Mech 2003; 70: 2473-2487.

[38] Codrington J, Kotousov A. A crack closure model of fatigue crack growth in plates of finite thickness under small-scale yielding conditions. Mech Mater 2009; 41: 165-173.

[39] Suresh S. Fatigue of Materials. Cambridge University Press, Cambridge, 1991.

[40] Blom AF, Holm DK. An experimental and numerical study of crack closure. Eng Fract Mech 1985; 22: 997-1011.

[41]   Stoychev S, Kujawski D. Methods for crack opening load and crack tip shielding determination: a review. Fatigue Fract Eng Mater Struct 2003; 26: 1053-1067.

[42]   Skorupa M. Load interaction effects during fatigue crack growth under variable amplitude loading – a literature review. Part I: empirical trends. Fatigue Fract Eng Mater Struct 1998; 21: 987-1006.

[43]   Shuter DM, Geary W. The influence of specimen thickness on fatigue crack growth retardation following an overload. Int J Fatigue 1995; 17: 111-119.

[44]   Borrego LP, Ferreira JM, Pinho da Cruz JM, Costa JM. Evaluation of overload effects on fatigue crack growth and closure. Eng Fract Mech 2003; 70: 1379-1397.

[45]   Basquin OH. The exponential law of endurance tests, in Proceedings of the Annual Meeting of the American Society for Testing Materials, 1910; 625–630.

[46]   Fatemi A, Yang L. Cumulative fatigue damage and life prediction theories: A survey of the state of the art for homogeneous materials. Int J Fatigue 1998; 20: 9–34.

[47]   Paris PC, Erdogan F. A critical analysis of crack propagation laws. J Basic Eng 1963; 85: 528-534.

[48]   Dill HD, Saff CR. Spectrum crack growth prediction method based on crack surface displacement and contact analyses. Fatigue crack growth under spectrum loads, ASTM STP 1976; 595: 306-319.

[49]   Budiansky B, Hutchinson JW. Analysis of closure in fatigue crack growth. J Appl Mech 1978; 45: 267-276.

[50]   Führing H, Seeger T. Dugdale crack closure analysis of fatigue cracks under constant amplitude loading. Eng Fract Mech 1979; 11; 99-122.

[51]   Newman Jr JC. A crack closure model for predicting fatigue crack growth under aircraft spectrum loading. Methods and Models for Predicting Fatigue Crack Growth under Random Loading, ASTM STP 1981; 748: 53-84.

[52]   Newman Jr JC, Phillips EP, Swain MH. Fatigue-life prediction methodology using small-crack theory. Int J Fatigue 1999; 21: 109-119.

[53]   Solanki K, Daniewicz SR, Newman Jr JC. Finite element analysis of plasticity-induced fatigue crack closure: an overview. Eng Fract Mech 2004; 71: 149-171.

[54]   Wang GS, Blom AF. A strip model for fatigue crack growth predictions under general load conditions. Eng Fract Mech 1991; 40: 507-533.

[55]   Codrington J. On the effect of plate thickness on post-overload fatigue crack growth. Int J Fract 2009; 155: 93-99.

[56]   Codrington J, Kotousov A. Crack growth retardation following the application of an overload cycle using a strip-yield model. Eng Fract Mech 2009; 76: 1667-1682.

[57]   Nowell D. A boundary element model of plasticity-induced fatigue crack closure. Fatigue Fract Eng Mater Struct 1998; 21: 857-871.

[58]   Codrington J, Kotousov A. The distributed dislocation technique for calculating plasticity-induced crack closure in plates of finite thickness. Int J Fract 2007; 144: 285-295.

[59]   Park HB, Lee BW. Effect of specimen thickness on fatigue crack growth rate. Nucl Eng Des 2000; 197: 197-203.

[60] Shin CS, Hsu SH. On the mechanisms and behaviour of overload retardation in AISI 304 stainless steel. Int J Fatigue 1993; 15: 181-192.

[61] Bichler Ch, Pippan R. Effect of single overloads in ductile metals: A reconsideration. Eng Fract Mech 2007; 74: 1344-1359.

[62] Schijve J, Broek D. The results of a test programme based on a gust spectrum with variable amplitude loading. Airc Engng Aerosp Technol 1962; 34: 314-316.

[63] Suresh S. Crack growth retardation due to micro-roughness: a mechanism for overload effects in fatigue. Scripta Metall 1982; 16: 995-999.

[64] Suresh S. Micromechanisms of fatigue crack growth retardation following overloads. Eng Fract Mech 1983; 18: 577-593.

[65] Borrego LP, Ferreira JM, Costa JM. Fatigue crack growth and crack closure in an AlMgSi alloy. Fatigue Fract Eng Mater Struct 2001; 24: 255-265.

[66] Pommier S, de Freitas M. Effect on fatigue crack growth of interactions between overloads. Fatigue Fract Eng Mater Struct 2002; 25: 709-722.

[67] Thorpe TW, Scott PM, Rance A, Silvester D. Corrosion fatigue of BS 4360:50D structural steel in seawater. Int J Fatigue 1983; 5: 123-133.

[68] Leis BN. Displacement controlled fatigue crack growth in inelastic notch fields: implications for short cracks. Eng Fract Mech 1985; 22: 279-293.

[69] Shin CS, Smith RA. Fatigue crack growth from sharp notches. Int J Fatigue 1985; 7: 87-93.

[70] Benedetti M, Fontanari V, Lütjering G, Albrecht J. The effect of notch plasticity on the behaviour of fatigue cracks emanating from edge-notches in high-strength β-titanium alloys. Eng Fract Mech 2008; 75: 169-187.

[71] Taylor D, Knott JF. Fatigue crack propagation behaviour of short cracks; the effect of microstructure. Fatigue Fract Eng Mater Struct 1981; 4: 147-155.

[72] Suresh S, Ritchie RO. Propagation of short fatigue cracks. Int Metals Rev 1984; 29: 445-476.

[73] Liaw PK, Logsdon WA. Crack closure: an explanation for small fatigue crack growth behaviour. Eng Fract Mech 1985; 22: 115-121.

CHAPTER 6

Non-Destructive Evaluation Methods for Solid Rocket Motor Structural Health Monitoring

S.Y. Ho

Defence Science and Technology Organisation, P.O. Box 1500, Edinburgh, SA 5111, Australia.

Abstract: This chapter describes miniature stress sensor technology for monitoring the thermal stresses and ignition pressurization loads in solid rocket motors. The study was part of a larger international (TTCP) collaborative effort carried out from 1988 to 2002 to validate the instrumentation and analytical stress analysis and service life prediction methodologies for solid composite rocket motors, and thus establish improved, more reliable, cheaper and non-destructive capabilities for service life prediction and extension. Different motor configurations were used by the different countries in this collaborative program. The Australian effort, described in this chapter, used an end burning generic research motor (Pictor). The embedded transducers, in this end burning motor and in the different motor designs used by the other TTCP countries, were found to be stable in the temperature range used in the environmental testing program and gave consistent data during propellant cure, environmental testing and static firing of the motors. The rocket motor instrumentation and data reduction techniques were described. The data from the instrumented motors under various thermal storage loading conditions (multiple thermal cycling, shocking, accelerated ageing at elevated temperature) were used to validate the stresses and critical failure modes predicted by structural finite element modelling and a modified fracture mechanics approach for nonlinear viscoelastic materials. These studies verified the ability of the miniature bond stress sensors to detect cracking / damage in the propellant charge. The advancement in bond stress sensor technology was further used to investigate failure analysis of rocket motors under ignition pressurization conditions. Results from these studies demonstrated that the sensors are safe for static firing and could accurately measure pressure in different regions of the burning motor. The stress sensor data from this international collaborative program showed that the stress sensor technology could be used for real-time structural health monitoring of solid rocket motors to detect cracks and debonds in the propellant and to continuously monitor the extent of damage. The results from the instrumented Pictor motor verified and validated the thermal distributions, stress / strain states and regions of high propensity for crack propagation predicted by finite element modelling and fracture mechanics. The use of the instrumented motor data in probabilistic service life prediction methodologies and other NDE methods are also discussed.

1. BACKGROUND

Until recently, conventional service life programs for solid propellant rocket motors comprised of experimentally based methods such as qualification testing and surveillance by scheduled material properties tests [1-3]. These traditional service life assessment methods are wasteful and expensive, as they involve static and trial firings of a large sample of motors followed by motor dissection and mechanical properties measurements to validate the service life of remaining motors in the inventory.

Several newer methods are available today for predicting the service life of solid propellant rocket motors, e.g. Cumulative Damage and Probabilistic Service Life analyses [4-6], structural FEM (finite element modelling) [7,8], and using instrumented motors to monitor stresses during storage and operation [9-20]. The most appropriate method to use would depend on the failure mode of the motor. Regardless of the method used, however, accurate stress predictions, under the expected environmental and operational conditions, are critical in developing a predictive service life analysis capability. For example, the probabilistic service life analysis approach (likely to be the preferred service life prediction method in the future) takes into account the statistical variability of chemical and mechanical properties of the propellant and the deployment (includes storage and operational) temperature of the motor, and relies on the availability of real-time data of the service loads and statistical data of the mechanical properties from lot-acceptance tests, *etc.* to give meaningful and reliable predictions. Hence real-time techniques for monitoring the stresses and temperatures in solid rocket motors are critical to the development of more accurate and reliable non-destructive methods for service life analysis.

This chapter describes previous work conducted by the author that was part of a larger international TTCP (The Technical Cooperation Program) collaborative program, Technical Panel WTP-4 Propulsion Technology, KTA 4-14 and KTA 14-23, carried out from 1998 to 2002 [15-20] to validate the instrumentation and analytical stress analysis and service life prediction methodologies that were available at that time, and thus establish improved, more reliable, cheaper and non-destructive capabilities for service life prediction and extension. Under these collaborative efforts, the participants (Australia, Canada, the UK and the US) fielded solid motors that were instrumented and then subjected them to agreed programs of testing and analyses. Four motor designs were used: the Canadian CRV-7, the UK's Structural Test Motor (STM), an Australian research motor (Pictor) and a US technology demonstration motor. This database was used to evaluate the performance of the new miniature stress sensors as well as to compare the predicted and measured stress and strain at critical areas in the propellant grain and relating this to any observed failure / damage.

One of the outcomes of the TTCP KTA 4-14 collaborative effort was that the use of miniature stress sensors in solid rocket motors and analogues was a reliable method for measuring long term thermally induced stresses, i.e. they have the potential for structural health monitoring applications [15-19]. The data from this instrumentation was also used to verify the theoretical models [21] and analytical service life prediction methods, such as structural FEM, for solid rocket motors subjected to thermal storage environments. In a later part of this international collaborative program (KTA 4-23), the miniature stress sensor technology was used to investigate failure analysis of solid rocket motors during ignition pressurization.

1.1. Review of Stress Transducer Technology

The early instrumented motor programs in 1963- 1975 identified many problems with the transducers and application techniques [20]. Aerojet instrumented two full-scale Minuteman 3rd stage motors and found that these instruments were unstable and gave random spurious data. Static test firing of the early Poseidon motor, instrumented with large Hercules stress transducers, resulted in a catastrophic failure. The bondline failure caused by these stress gauges triggered initiation of detonation. The causes of the problem with these early instruments included the following:

- Thin diaphragms caused high structural interaction and viscoelastic transducer response;
- Gauge leakage, chemical corrosion and gauges / leadwire contacts;
- Unstable / low accuracy electronics;
- Gauges were adequate for pressure but not thermal tests.

The improved gauges developed in the period 1975-1983 were too cumbersome for motor use. These through-case gauges worked well in steel analog motors. Screening tests were needed to eliminate poor stability units.

In 1984-1987 low-profile normal stress gauges were developed and these worked well, as demonstrated in Trident II static and flight motors [20]. The shear gauges, however, had a problem with the potting material becoming stiff at temperatures below 0°C. Screening tests with these miniature gauges showed gross instability, and the need for stabilized miniature stress gauges for termination regions became apparent.

From 1988 – 2002, a third generation of miniature bondline stress sensors were fabricated which have increased accuracy and stability, and could be used for in-situ health monitoring of tactical motors [22-23]. These sensors were utilised in the instrumented motors studied in the TTCP collaborative efforts (KTA 4-14 and 4-23) mentioned in Section 1. These stress sensors met the long term stability (critical for embedded solid propellant transducers) and the chemical protection requirements for the electrical components from the corrosive propellant environment. The transducer features used to ensure accuracy and stability are listed in references [22] and [23].

Micron Instruments, through US Air Force and UK Defence Science and Technology Research funding, refined the technology to fabricate a stable, high signal-to-noise, miniature normal stress sensor using semi-conductor strain gauges. These Micron miniature stress sensors [13, 23] have a low flat miniature profile with a diameter of 7.6 mm and 2.0 mm thickness, so that they are more easily installed in rocket motors with minimal effect on the grain stress distribution. They exceeded the design goals for thermal hysteresis and long term stability:

- Thermal hysteresis was less than 3 kPa (0.5 psi) (the design goal was 7 kPa (1 psi) when thermally cycled between -45°C to +65°C);

- The average drift of the sensors was less than 5 kPa per year with many sensors drifting less than 2 kPa per year (long term stability design goal was 7 kPa per year).

Micron Instruments [23] has also developed a normal bond stress / temperature dual sensor (DBST) [23] with the dual reading capability which allows temperature and bond stress data to be obtained simultaneously. A silicon temperature element the same size as the strain gauge, which has no significant strain sensitivity and not affected by pressure or case induced strains, was developed and bonded directly to the stress sensor diaphragm. Since this study in the late 1990's, Micron Instruments has made many improvements to their miniature dual measurement sensor. The specification of Micron's DBST sensors, optimized for measuring rocket motor propellant bond stress and temperature at the inside surface of the motor case wall, can be found in reference [13]. The latest DBSTs do not exhibit hysteresis problems anymore because Micron Instruments has developed a way to match strain gauges before installing them into the sensor body [34]. The other improvements Micron Instruments made to the DBSTs include wireless DBSTs [35]. Besides the obvious benefits offered by wireless technology, the wireless DBSTs would eliminate problems associated with stress concentration in the case or propellant region generated by wiring, and avoid some of the difficulties with processing and protecting the wirings from corrosion, *etc.*

In the mid 2000's, with the successful development of the normal stress sensor based on the Micron design, there was renewed interest to develop a shear stress sensor. Efforts to develop shear stress sensors in the past have to be abandoned for a number of reasons, including low shear sensitivity, unacceptable zero-offset stability and large in-situ shear stress sensors were not accepted by the rocket community due to safety concerns. Reference [26] describes a finite element modelling effort to determine if the diaphragm deformations of a shear stress sensor design based on the Micron bond stress sensor housing could be used to measure shear at the bondline. The study showed the locations for placement of the strain gauges on the diaphragm for measuring radial strains that can be analysed to deduce shear stress magnitude and direction.

2. ROCKET MOTOR INSTRUMENTATION

2.1. Test Motor Design and Processing

A generic research motor with an end burning design, Pictor (Fig. **1a**), was used for this study. It is filled with a non-aluminized HTPB/AP propellant, inhibited with a tapered thickness beaker (inhibitor) of adiprene/TMP. This inhibited charge is case bonded at the head end with an epikote/versamid adhesive. The motor has a maraging steel case and is thermally insulated with EPDM which is spun cast in the case. There is a small air gap in the side wall of the motor between the inhibitor charge and the insulated case.

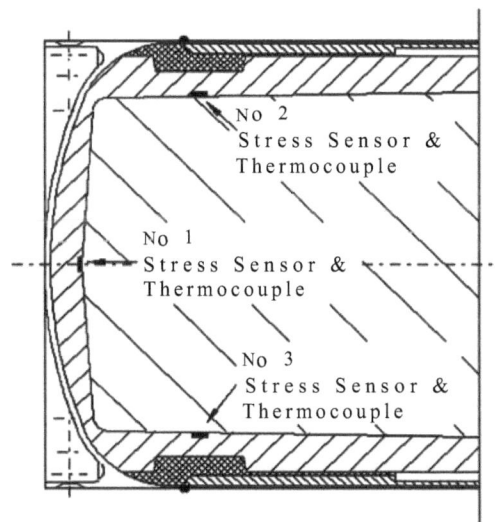

Figure 1a: Instrumented Pictor motor, as taken from reference [7].

Instrumentation of the Pictor motor involved the following steps:

1.   Bonding the sensors into the inhibitor;

2.   Casting the propellant into the inhibitor and

3.   Bonding the charge into the steel case and curing the adhesive used in the case bonding.

The processing steps are shown in Fig. **2**. During these steps, the response of the sensors were measured to (1) obtain zero offset correction factors of the sensors after mounting, (2) correct for cure shrinkage and mechanical loads due to the casting corset, and (3) correct for shift due to mechanical load from case bonding.

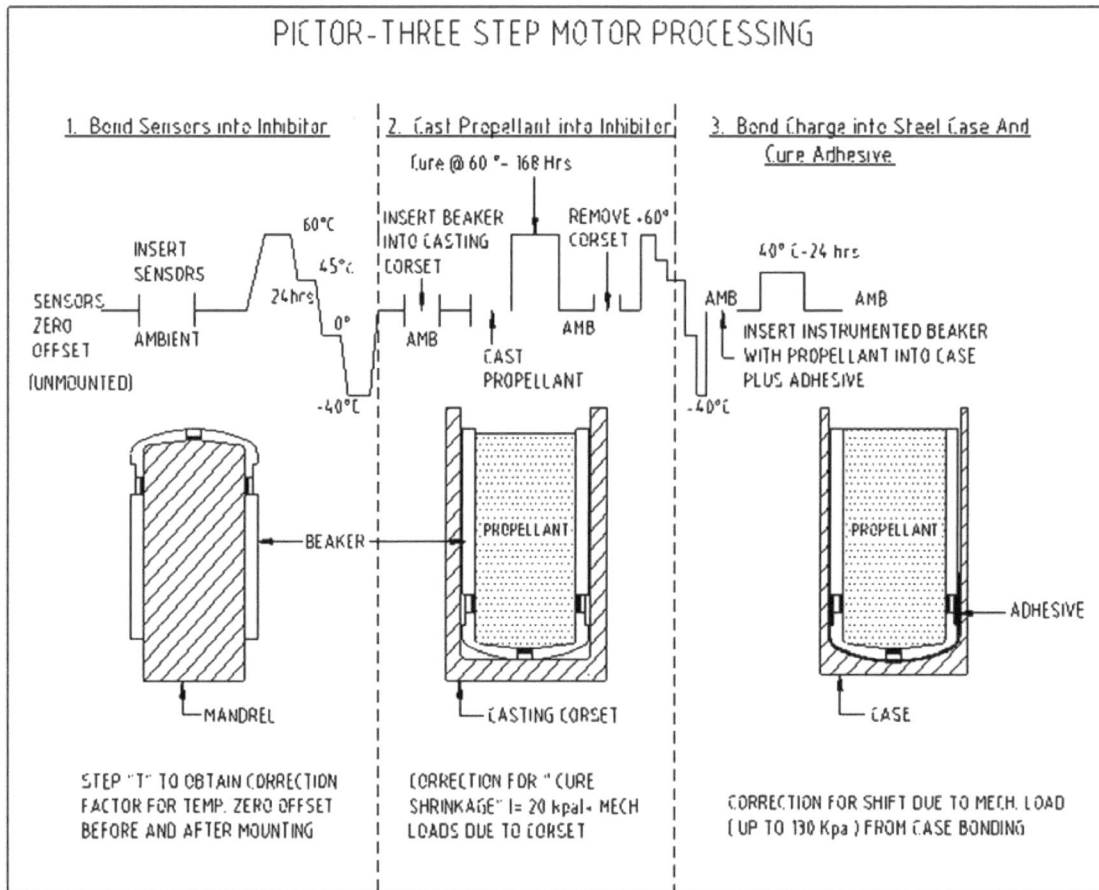

Figure 2: Pictor motor processing, as taken from reference [11].

2.2. Gauge Installation and Data Acquisition

Environmental Testing

For the environmental testing studies, the motors were instrumented with Sensometrics (Simi Valley, CA) miniature stress sensors (see Fig. **3**). These sensors (5 mm diameter by 1.5 mm thick) were used with 5 milliamp constant current excitation. The sensors were flush mountable with a very stiff diaphragm, to minimize their effect on the stress distribution of the rocket motor grain. They were factory calibrated from 0-100 psi and accurate to 1% full scale with 5 ma of constant excitation from the power supplies. Thermal hysteresis testing and cooling cycles between 160 and -65°F indicated drifts between 100-300 µV (ca. 1-5 psi of measurement error) could be expected. The sensors with the lowest hysteresis delta figures were selected.

The sensors (3 per motor) and thermocouples were installed near the regions of critical stress identified by structural FEM, i.e., at the propellant/inhibitor interface adjacent to the holdback groove of the inhibitor and also at the head end of the motor (see Fig. **1a**). The positions of sensors 2 and 3 are identical in order to determine the variation in the measurements due to the sensors. The inhibitor beaker was machined with 18 mm holes and side grooves (6 mm wide) for the sensor and thermocouple wiring up to the nozzle end of the motor. A plug and base plate were manufactured from the beaker material and the sensor bonded to the base plate with a fast setting epoxy adhesive after degreasing the bonding surfaces. Thermocouples (type T copper/constantan) were inserted into the holes drilled in the plugs, located at the sensor case, and filled with epoxy.

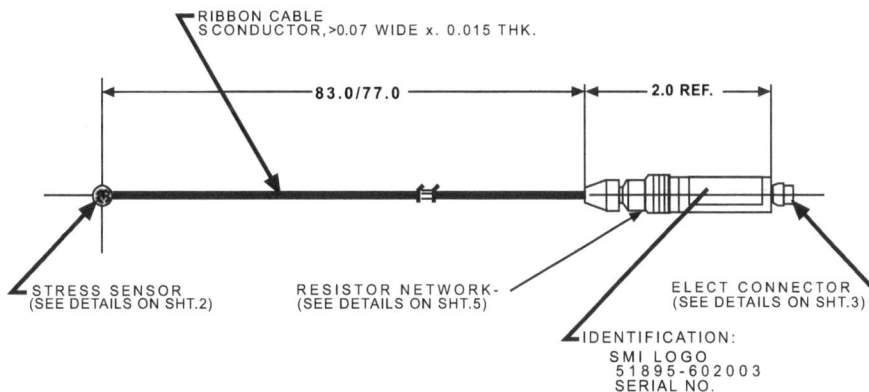

Figure 3: Miniature normal stress sensor (Sensometrics type 602003, dimensions: 5mm diameter x 1.5 mm thick).

Ignition Pressurisation

For the ignition pressurization tests, normal bond stress miniature sensors from Micron Instruments (MICRON 140480 specification) were used. The instrumented Pictor motor for this study is illustrated in Fig. **1b**.

The sensor cable exit was through the head end (instead of the usual nozzle end for thermal load tests) to ensure integrity of the measurement throughout the motor burn time. Exit through the rear would have resulted in early damage to wiring (i.e. during the ignition phase) due to the difficulty of shielding from the high temperature and pressure of motor exhaust with this nozzle design. A 6mm hole (the minimum needed to meet cable size exit requirements) was drilled in the head end. This was expected to be a low pressure area (due to partial case bonding between the key area and the head end) and if so, it would avoid the need for a high pressure electrical connector to be externally welded to the case around the hole. A preliminary test using an inert charge cast into a representative motor case was carried out to test this premise. Pressure was applied via a sealed nozzle enclosure and pressure pipe from Nitrogen gas cylinders and increased to a level above the typical measured chamber pressure. There was no evidence of structural failure at the head end and nor was there any evidence of extrusion of case bond material through the exit hole area.

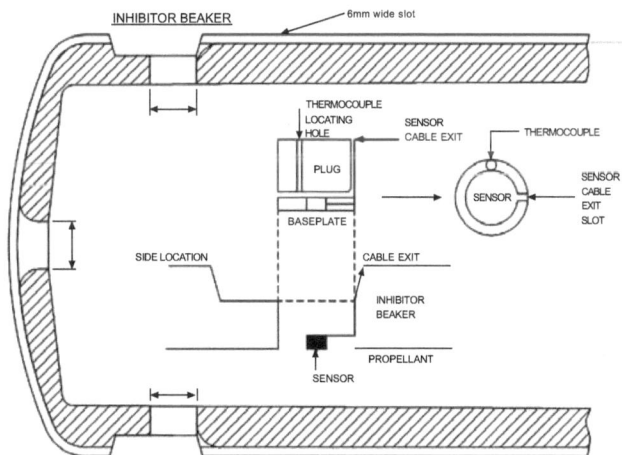

Figure 1b: Instrumented Pictor motor for ignition pressurization tests, as taken from reference [11].

All data was collected on a Datataker 500 data logger with 10 differential input channels. The stress sensors were connected to the data logger via SMI 400216 shielded cable and a K18 constant power supply. Excitation for the sensor was a 5 ma constant current and is measured as a voltage drop across a 10 Ω precision resistor.

2.3. Selection / Installation of Sensors

To accurately measure stresses in solid propellant rocket motors, sensors with the following features were required.

1.   Long term mechanical and electrical stability. Any electrical drift above a suitable level (1% full scale per year) would be incorrectly interpreted as stress changes.

2.   Adequate chemical protection of parts from the corrosive propellant environment.

3.   Small enough to be recessed into the beaker so that the sensing face is flush with the propellant to minimize the stress disturbance factor.

The design of the MICRON standard 689 kPa (100 psi) steel based sensors is conservative and has more than 30 times over the pressure capability which will also allow it to be used to measure chamber pressures up to 21 MPa (3000 psi) if required. These sensors are bonded inside the motor case and can be used to obtain the reactive pressure profile of the motor during operation until the flame front reaches the sensor.

2.3.1. Thermal Hysteresis and Zero Offsets Measurements

A critical requirement for the stress sensor is very stable zero offset and pressure sensitivity. While pressure sensitivity does not change significantly with time, the zero offset of these transducers may drift with age and also is temperature dependent (because of their semi-conductor properties). The sensors were factory calibrated for pressure sensitivity (mV/psi) and then screened for electrical stability by temperature cycling with no load output (i.e. the transducer output is influenced only by thermal loads) at various temperatures.

The transducer no load electrical output (zero-offset) with temperature was measured before and after installation in the inhibitor beaker (i.e. prior to propellant casting) using conditioning chambers. The temperature zero offset is not regarded as an error but is repeatable and unique for each transducer. The ZO(T) value after installation is regarded as a preload stress and may be either tension or compression.

The hysteresis and temperature effects of the stress sensors were determined by temperature cycling the instrumented beaker from -40 to +60°C. Temperature zero offset coefficients were then obtained from fourth order polynomial fit to each set of data for the different sensors. In most cases, the shape of the zero offset vs. temperature curve changed when the stress sensor was mounted in the beaker (see for example Fig. **4**).

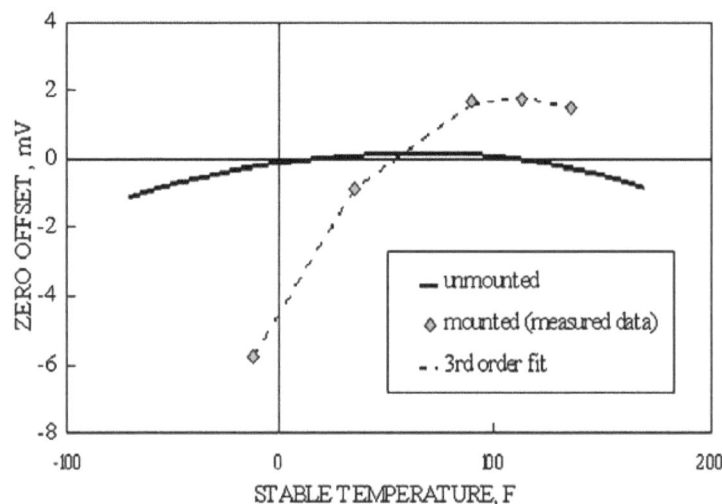

Figure 4: Instrumented Pictor inhibitor beaker – zero offset response of a stress sensor (A540), as taken from reference [7].

Examples of the thermal hysteresis data, supplied by the manufacturer, for the stress sensors used in the ignition pressurization experiments are shown in Appendix 1. In all cases, the sensors showed thermal hysteresis that was less than 2 psi when cycled in the temperature range -40 to +60°C. Note that the sensors should have a hysteresis response that is close to 1 psi for thermal cycling studies. For the ignition pressurization studies, however, this is not necessary as the induced stresses are much higher.

Pressure Sensitivity

For the ignition pressurization tests, normal bond stress sensors (based on the MICRON 140480 specification for the 17-4 PH steel sensors) that could be calibrated to a value above the expected ignition pressures were selected. The pressure sensitivity of the transducers was measured using an oil filled dead weight tester with an oil seal adaptor to avoid contamination of the sensing face.

Data Reduction

Measured stresses were calculated using equation (1) below:

$$\sigma_p = \left[(OP)(0.5/Ex_t) - ZO(T)\right]\left[1/S_e(T)\right]I_f D_f + C + (B - 14.7) \tag{1}$$

where

$0.5/Ex_t$ = transducer excitation power fluctuation correlation,

$ZO(T)$ = transducer zero offset vs. temperature behaviour (mV),

$\qquad = Z_0 + Z_1 T + Z_2 T^2 + Z_3 T^3 + Z_4 T^4,$

$S_e(T)$ = transducer sensitivity vs. temperature behaviour (-mV/psi),

$\qquad = S_0 + S_1 T + S_2 T^2$ (the default value for S_1 and $S_2 = 0$ was used),

C = cure fluid weight correction (psi),

T = temperature (°F),

I_f = transducer-propellant interaction factor (correction is small because sensor is flush and has been neglected),

D_f = transducer stress disturbance factor (correction is small and has been neglected).

In the initial stress calculations corrections have been made for the temperature zero offset of the transducers, obtained from the fourth order polynomial fit. However, corrections for the transducer-propellant interaction and stress disturbance of the gauges are insignificant (as indicated by the 2D-viscoelastic finite element model with the stress sensors) and have been neglected.

Since the Pictor motor processing involves a number of steps (see Fig. **2**), additional measurements were made during the multiple step processing. These shifts were relatively small (up to 20 kPa), except for the shift due to mechanical load from case bonding (ca. 100 kPa).

Measured stresses from the two sensors at the side of the Pictor charge taken during cure is shown in Fig. **5**. Both gauges showed the same trend and the measured stress responses were consistent during cure. The stresses gradually increased during the cure process (propellant changes from liquid to solid) but final stresses were very small. A stress free temperature of 55°C was determined from the stress measurements.

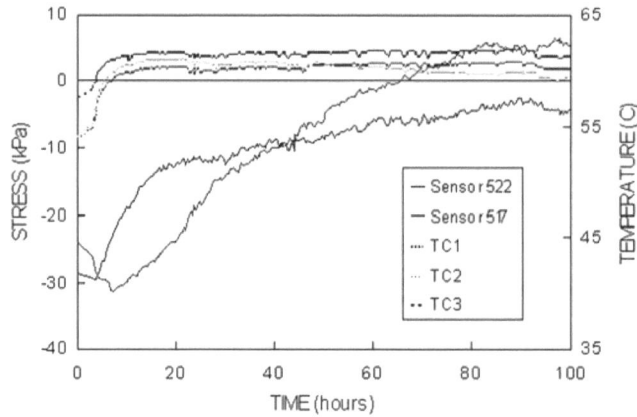

Figure 5: Pictor charge – cure at 60°C, as taken from reference [9].

The initial cooldown stresses for the Pictor motor at 50°C is shown in Fig. **6**. The stresses were approximately 60 kPa.

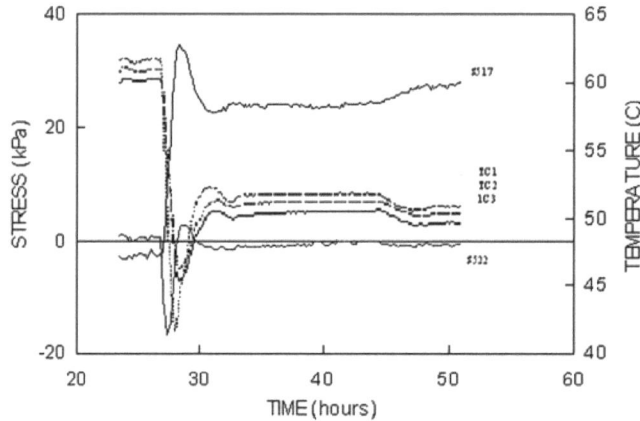

Figure 6: Pictor charge – cooldown at 50°C, as taken from reference [9].

Thermal cooldown tests were conducted to determine the residual stresses at different temperatures [7]. Fig. **7** shows the equilibrium stresses as the charge was allowed to cool down from 60°C after the propellant had cured. A stress free temperature of 55°C was determined.

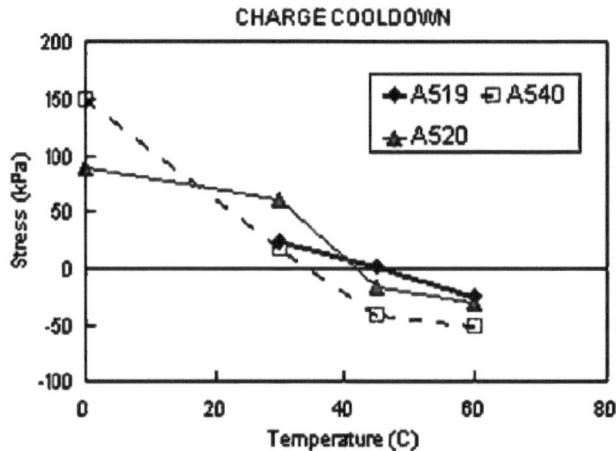

Figure 7: Residual stresses at different temperatures, as taken from reference [10]

2.5. Test Results from Instrumented Motors

2.5.1. Environmental Testing

For the environmental testing, one instrumented (I) and two uninstrumented (UI) motors were used for each of the tests presented in Table **1**. The motors were to be cooled to ambient temperature and then under one of the following tests: thermal shock, thermal cycling, accelerated ageing or ambient ageing.

Table 1: Environmental testing program.

Number of Motors		Thermal Environment
1I (Pictor 2), 2UI	Thermal shock	16 hours at -40°C and 16 hours at +50°C.
1I, 2UI	Thermal cycle	Minimum of 5 cycles of 30 days at 45°C and 20 days at -30°C.
1I, 2UI 2 sealed charges	Accelerated ageing	Minimum of 64 weeks at +60°C.
1I (Pictor 4), 2UI	Ambient ageing	Storage at 22°C for at least 2 years.

Thermal Shock

The Pictor 2 motor was subjected to 13 thermal shock cycles. The bondline stresses and thermocouple temperatures for the heating and cooling cycles are shown in Figs. 8a and b. The two sensors at the side of the motor showed the same trend. Relaxation was also observed and was more apparent in the cooling cycles. For both the heating and cooling cycles, the stresses decreased with each thermal shock cycle due to mechanical damage which was evident as a reduced compressive or tensile stress. In the cold cycle, the stresses decreased (95% reduction) very markedly after the 4th cycle, whereas in the hot cycle the damage was not apparent until the tenth cycle.

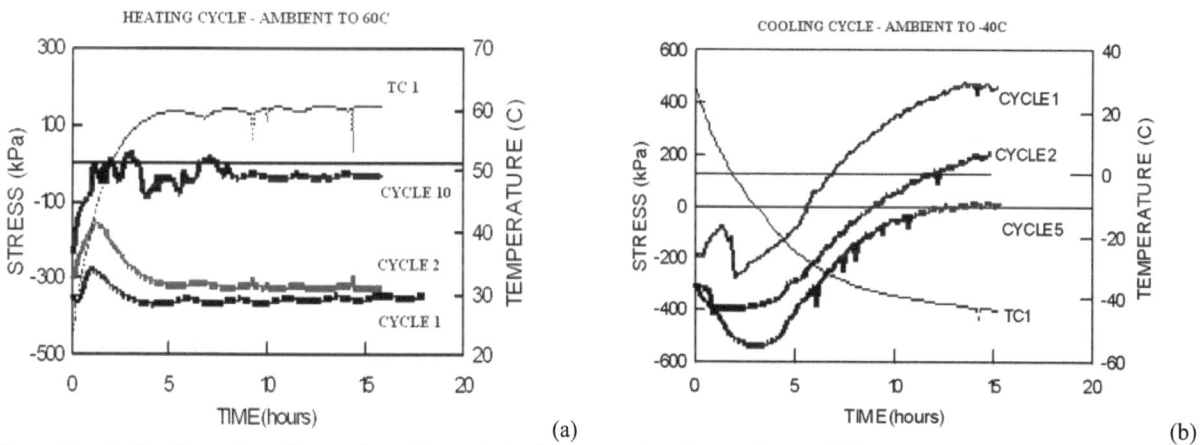

Figure 8: a, b: Heating and cooling cycles of thermal shock test, as taken from reference [10].

The reduction in the measured stresses may be due to debonding at the propellant-inhibitor interface. This was confirmed by radiographs (the motors were X-rayed after each hot and cold cycle) which showed a dark line, indicating a separation or debond, at the propellant/inhibitor interface adjacent to the holdback groove in the inhibitor, and is consistent with the critical stress areas predicted by FE models (see Section 3.2). The two uninstrumented motors also showed cracking in the propellant near the head end and adjacent to the holdback groove of the inhibitor after the first and second thermal shock cycles.

Ambient Ageing

Fig. **9** shows the bondline stresses of the Pictor 4 motor during ambient ageing over a period of 250 days. The two stress sensors at the side of the motor gave similar measurements. At the start of the test the measured stresses were 70 kPa when the temperature was 15°C. After 5 days of storage at ambient temperature (22°C) the stresses decreased to 50 kPa and remained at this level during the 280 days of storage.

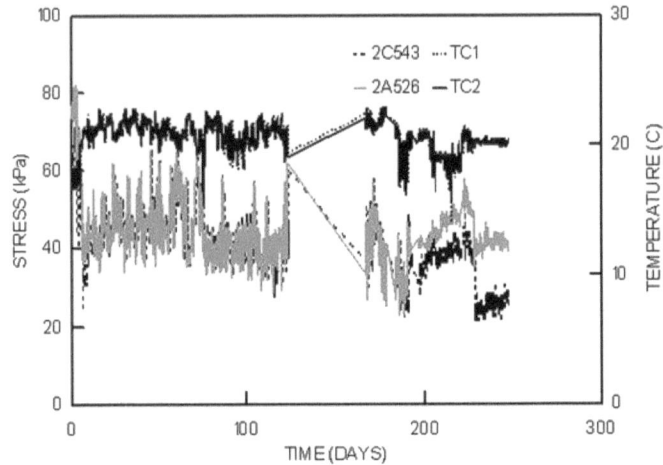

Figure 9: Ambient Ageing Test, as taken from reference [9].

Thermal Cycling

Fig. **10** shows the bondline stresses for the thermal cycling test after ten cycles of 30 days at +45°C and 20 days at -30°C. The equilibrium stresses from the sensors at the side of the motor were -10 and -92 kPa for the hot cycle and 305 and 105 kPa for the cold cycle. There were no measured changes in the bondline stresses after 5 complete thermal cycles. The uninstrumented motors have been subjected to 16 complete cycles and were radiographed at the end of each cold cycle. No cracks / damage were evident in the radiographs.

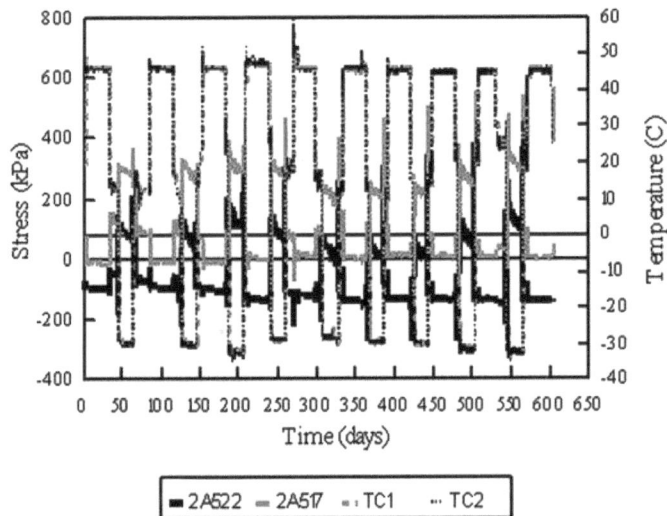

Figure 10: Thermal cycling test results, as taken from reference [9].

Accelerated Ageing

Fig. **11** shows the bondline stresses for the accelerated ageing test up to 300 days. Initially, the stresses from the two sensors at the side of the motor were around -50 kPa and changed to zero after ca. 10 days of ageing at 60°C. This indicates that the motor zero stress temperature is shifting to a higher temperature. This is consistent with other previous work [14].

Ignition Pressurisation

Three Pictor motors were instrumented with the MICRON14080 miniature normal stress sensors (3 per motor, calibrated for pressures up to the expected chamber pressure of the motor, and T-type thermocouples. The test program for motor fabrication, instrumentation, thermal shock and static test firing is summarized in Table **2**. For this study, three motors were subjected to very severe thermal shock conditions to induce cracks in the solid propellant, in order to test the capacity of the sensors to detect cracks / damage in the motor.

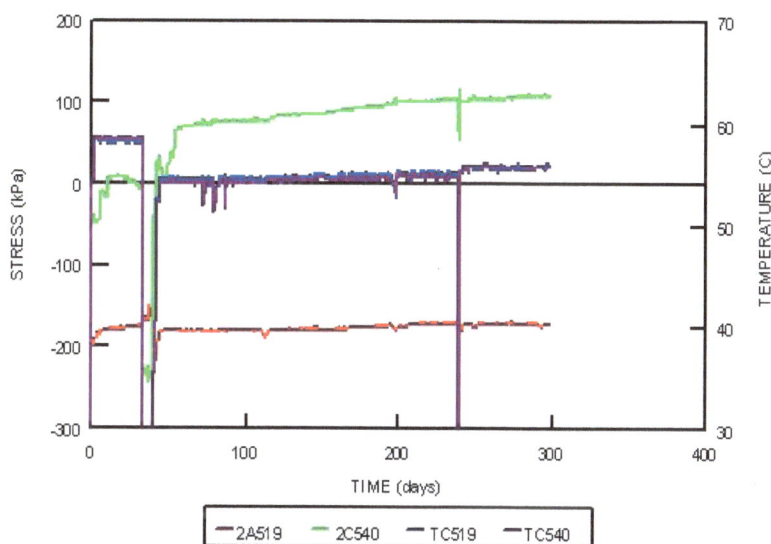

Figure 11: Accelerated ageing test results, as taken from reference [9].

Table 2: Ignition pressurization studies - test program for instrumented motors.

Thermal shock:

2-3 motors - induce crack(s) in bondline near head end of motor.
Thermal shock conditions: several cycles of 8 hrs. at
–40°C and 8 hrs at 60°C.
X-ray motors after each thermal shock cycle and before static firing.

Static firing:

2 instrumented motors (no cracks) at 20°C. (Pictor 16 and 17).
1 instrumented thermal shocked motor (with crack) at 20°C.
1 instrumented thermal shocked motor (with crack) at 60°C.
1 instrumented thermal shocked motor (with crack) at -40°C.

The chamber pressure, bondline stresses from the stress sensors, and temperatures from the embedded thermocouples vs. time profiles for one of the instrumented motors (Pictor 17) statically fired at 20°C are shown in Figs. **12** and **13**. The results from the second instrumented motor (Pictor16) statically fired at the same temperature were similar. The temperature in the vicinity of the stress sensors did not rise till the burning surface approached the sensors. The bondline stresses measured from the sensors at the head end and side of the motor are in compression during motor operation and follow very closely the chamber pressure vs. time profile.

Data obtained from the static firings demonstrated the ability of the bond sensors to accurately measure the pressure in different regions of the burning motor.

The sensors are picking up the hydrostatic pressure from ignition instantaneously. These results indicate that the high strain-rate during ignition pressurisation shifts the mechanical behaviour towards the short time region and the propellant grain exhibits linear elastic behaviour, i.e. the induced maximum principal stresses in the propellant are similar to the applied pressure load. The contributions from thermal loading, viscoelastic effects and short, non-propagating cracks are small in comparison with the ignition pressurisation load and may not be manifested in the measured bondline stresses.

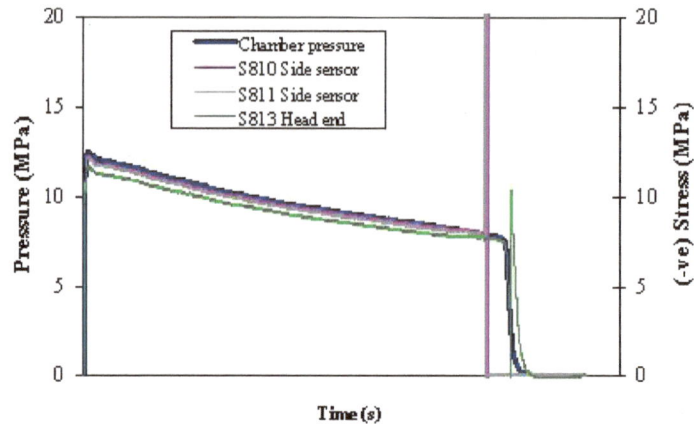

Figure 12: Plot of pressure and stress vs. time. Ignition pressurisation (static firing at 20°C) test results for Pictor motor 17, as taken from reference [11].

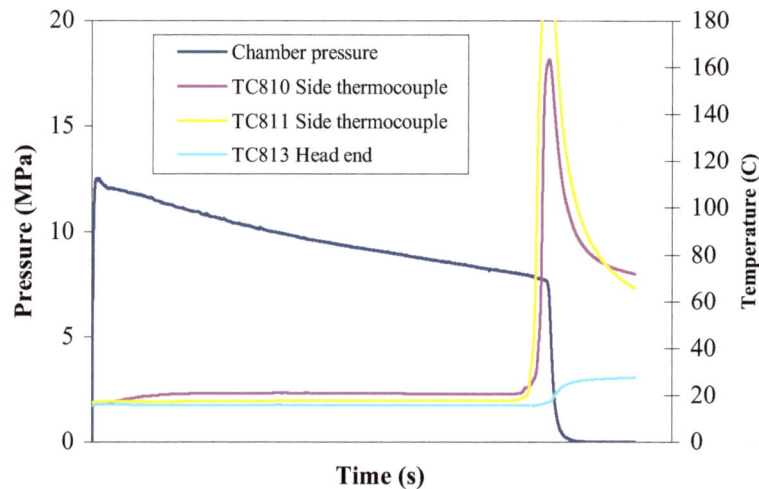

Figure 13: Plot of pressure and temperature vs. time. Ignition pressurisation (static firing at 20°C) test results for Pictor motor 17, as taken from reference [11].

3. VALIDATING THERMAL, STRESS AND FRACTURE ANALYSES RESULTS FROM FEM

The measured stress and temperature data from the instrumented Pictor motors were used to validate the analytical thermal, stress and fracture analyses from FEM. The development of non-linear viscoelastic (NLVE) and inelastic fracture mechanics capabilities in the STRAND finite element code has been described in detail in reference [8]. Modifications were also made to this code to model combined thermal and pressure loading. The stresses predicted by this NLVE method under various thermal loading conditions compared reasonably well with the stresses measured using the miniature stress sensors embedded in the end burning motor [8]. Only the results for thermal shock and ignition pressurisation load cases are presented below.

3.1. Stress Analysis

A 2-D viscoelastic analysis of the Pictor motor with embedded stress gauges was conducted for the upward (ambient to +47°C) and downward (+47 to -23°C) thermal transients in the thermal shock test. The von Mises stress contour predicted by the non-linear viscoelastic axisymmetric model for the upward thermal transient is shown in Fig. **14**. The solutions assume that the motor is in a state of zero stress and strain at the start of the transient. Hence, the predicted stresses would have to be corrected for residual stresses present in the motor at the starting equilibrium temperature, particularly if the difference between the starting temperature and the temperature in the stress free state is large.

The model shows that the high stresses in the propellant grain are concentrated at the head end of the motor and also at the side adjacent to the holdback key. In the upward thermal transient, the predicted von Mises stress in the propellant adjacent to the holdback key is around 130 kPa. This is in agreement with the experimental values of 133 and 80 kPa. In the downward thermal transient, the stress in the propellant adjacent to the holdback key is around 400 kPa. This compares with the experimental values of 500 and 380 kPa.

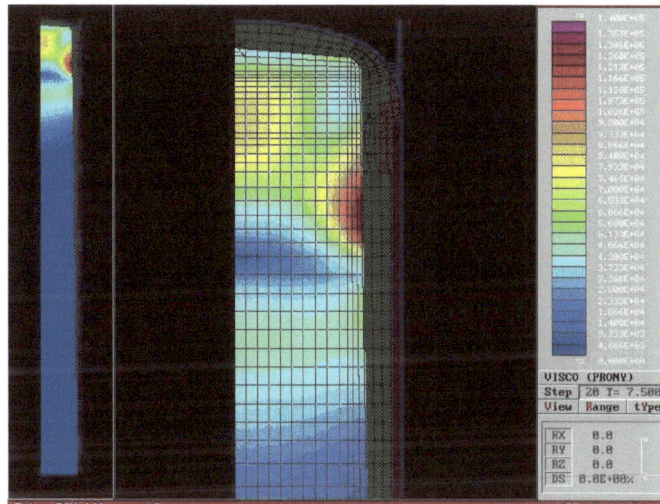

Figure 14: Von Mises Stress Contour, as taken from reference [9].

The analysis also showed that the air gap, being a poor conductor of heat, has a significant effect on the predicted stress distribution. During the temperature transient the propellant and the inhibitor expand into the air gap and in some cases the air gap opens up. This has a significant effect on the magnitude and distribution of the stresses and strains during the transient.

Fig. **15** shows the deformed shape of the strain contours in the motor at the end of the upward thermal transient. It shows the propellant expanding and compressing against the inhibitor. The principal strain in the propellant in the critical regions is 1.3 - 1.5%.

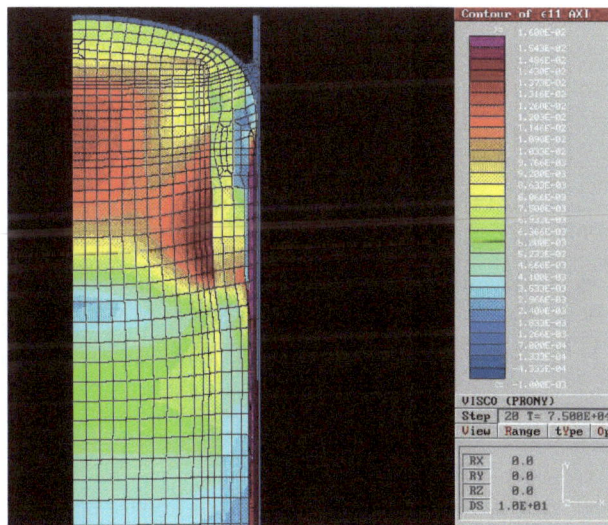

Figure 15: Principal strain contour, as taken from reference [9].

3.2. Thermal Analysis

The temperature distribution in a 3D model of Pictor for the upward transient, from a FE thermal analysis, is illustrated in Fig.16. The motor was equilibrated at ambient temperature and step change to 47°C. It predicts that after 2 hours, the temperature differences between the motor case and propellant at the two different gauge positions, i.e. at the head end and at the side adjacent to the holdback groove of the inhibitor, are 14 and 19°C respectively, as observed in the experiment. The model also correctly predicts that thermal equilibrium is reached after 21 hours for the downward thermal transient (see Figs. 17a and b).

A comparison of the results from the thermocouples in the instrumented motors to the temperatures predicted by the FE thermal analysis are shown in Figs. 17a and 17b for the downward thermal transient. The effect of the convective heat transfer and emissivity coefficients, the key factors that influence the closeness of the fit between the finite element and the thermocouple results, are illustrated. Errors in the convection coefficient seem to be less significant than those in the radiation coefficient, as can be seen in the results from the FE analysis with different heat transfer coefficients (Figs. 17a and 17b).

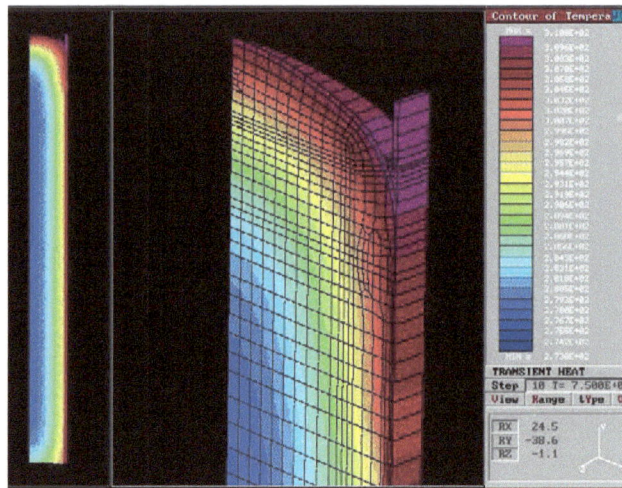

Figure 16: Temperature distribution for heating cycle of thermal shock test, as taken from reference [9].

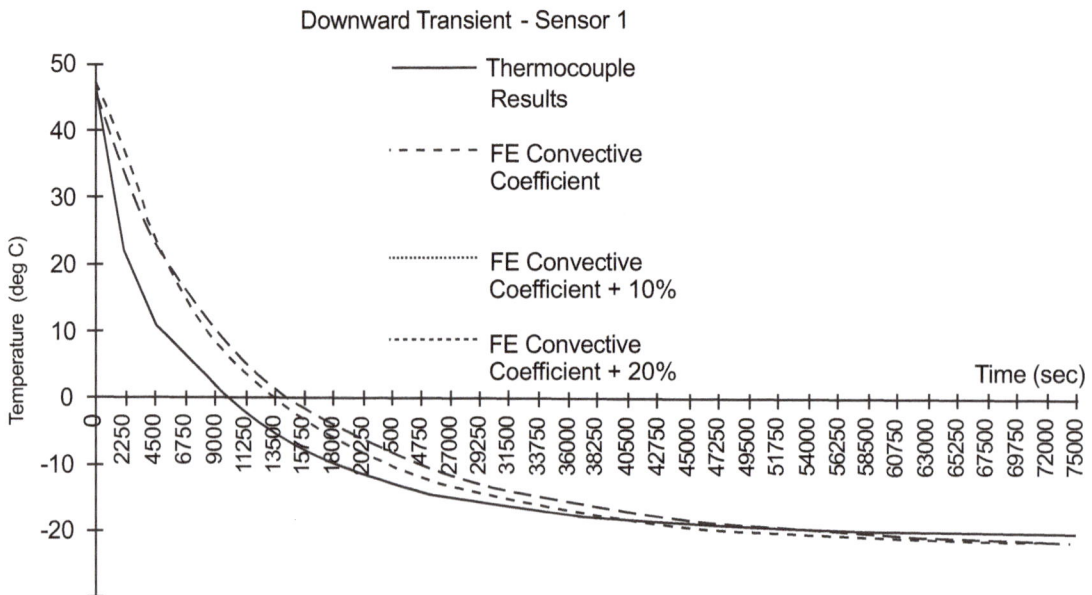

Figure 17a: Comparison of thermocouple temperatures measured from instrumented motors with predicted temperatures from finite element thermal analysis for the downward thermal transient from +47°C to -23°C. FE results illustrate the effect of convective coefficient on temperature transient.

Downward Transient - Sensor 1

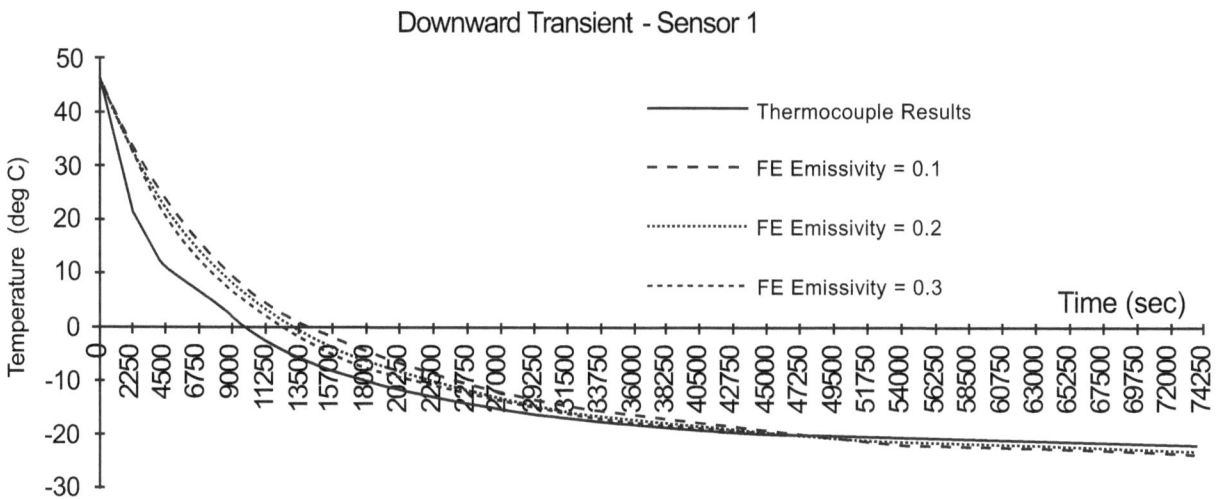

Figure 17b: Comparison of thermocouple temperatures measured from instrumented motors with predicted temperatures from finite element thermal analysis for the downward thermal transient from +47°C to -23°C. FE results illustrate the effect of emissivity on temperature transient.

3.3. Fracture Analysis

The results from the thermal shock tests (see Fig. **8**) clearly showed that the miniature stress sensors can detect cracking / damage in the propellant and propellant-inhibitor bondline. A reduction in the bondline stresses was apparent during the fifth cold cycle and the location of the debond was confirmed by radiographs taken after the thermal shock cycle.

These results are also consistent with the regions of high propensity for crack propagation predicted by fracture mechanics. The critical strain energy release rate, G_c, at various temperatures and strain-rates were determined using a modified fracture mechanics approach [8] that takes into account bulk inelastic behaviour. This fracture analysis theory was implemented in our structural FEM code and provides a capability to predict the location of the crack and the minimum length above which the crack will propagate in the motor [8].

The fracture resistance of the propellant was characterised using single-edged notched specimens. This fracture energy based failure criterion has been incorporated into our viscoelastic finite element code to predict the propensity of crack growth during a thermal transient.

The changes in the propensity for crack propagation during one complete thermal shock cycle are shown in Fig. **18**. The model predicts that the region of highest propensity for crack propagation is at the corner of the dome in the propellant charge. The critical crack lengths required for propagation (i.e. cracks will not propagate below this minimum value) are: 3-10 mm in the hot cycle and 0.2-2 mm in the cold cycle. On equilibration to 25°C, the propensity for crack propagation decreases and the critical crack lengths are longer. During the hot cycle at 60°C, the critical crack lengths decrease again but are longer than those at -40°C, i.e. the propensity for crack propagation is much higher at -40°C, where the motor encounters higher thermal tensile stresses.

3.4. Ignition Pressurisation

The viscoelastic solutions were started at a steady-state ambient temperature of 55°C (cure temperature), which is the stress free temperature of the motor, and equilibrated at 20°C or 60°C (i.e. the static firing temperature) before applying the pressure loading. It is assumed that the temperature in the propellant grain does not change significantly during the burn time of the motor, which is relatively short compared to the time taken for heat transfer from the burning propellant surface. This is supported by the static firing data from the instrumented motors (see Fig. **13**). The pressure loading used in the analysis is the peak pressure measured during static test firing. Grain-burnback (estimated from the burning rate of the propellant) was modelled by manually removing elements from the model.

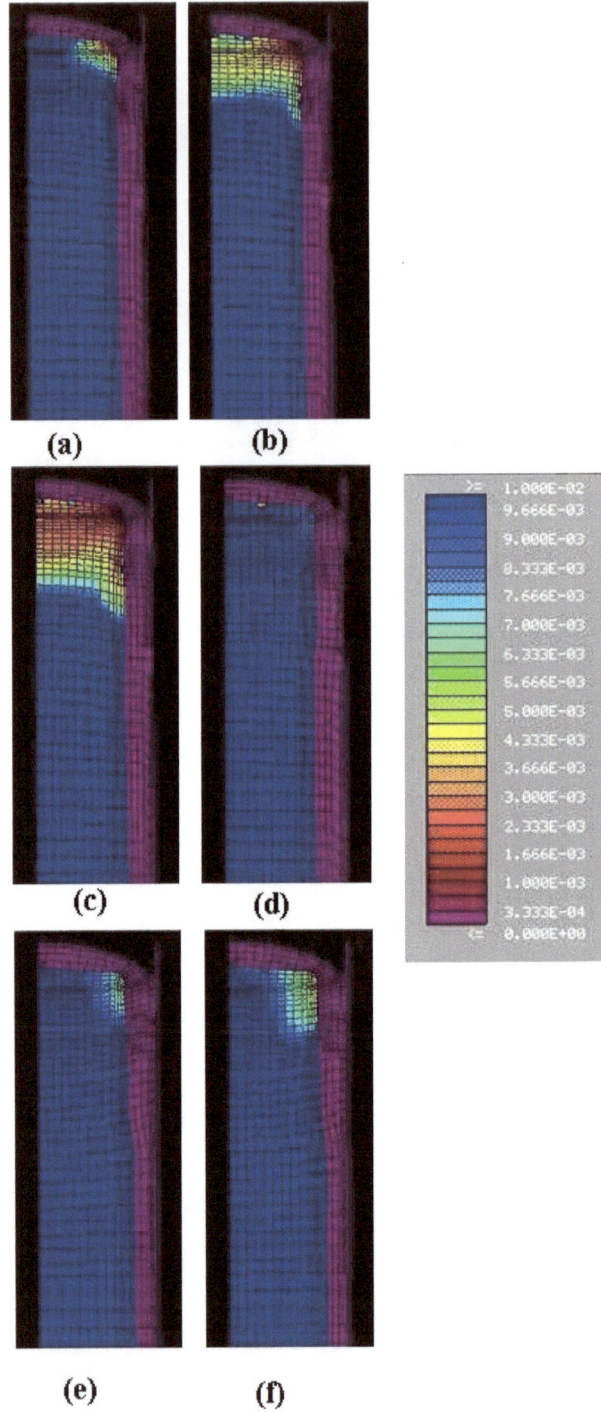

Figure 18: Contours of propensity for crack propagation during thermal shock cycle after (a) 1.3 hr., beginning of cold cycle at –40°C; (b) 2.7 hr., -40°C; (c) 8.7 hr., -40°C, (d) 22.7 hr., end of equilibration at 25°C for 8 hr., (e) 32 hr., 60°C; and (f) 40 hr., end of hot cycle at 60°C, as taken from reference[8].

Pressure boundary conditions were applied on the propellant burning surface. A linear static analysis was also conducted. There was little difference in the results between the linear static and non-linear viscoelastic analyses, because the effects from non-linear viscoelasticity are small in comparison with the effects of the high surface pressure load. The predicted radial stresses for the head end and side sensors were consistent with the measured bondline stresses from the instrumented motors.

4. APPLICATION OF DATA FROM INSTRUMENTED ROCKET MOTORS FOR SERVICE LIFE DETERMINATION

During the course of the TTCP collaborative programs, new methods of service life prediction that utilised the data from the instrumented motors emerged. As well as using the instrumented motors for (1) structural health monitoring to give a real-time indication of when cracks and /or failure occur, (2) validate theoretical viscoelastic models and stress and strains predicted by structural FEM, the data could also be used for (3) service life determination using the deterministic approach and probabilistic methods.

A method for service life determination using the deterministic structural analysis approach is discussed in Chapter 7 of this book. Reference [4] describes a probabilistic service life method using the instrumented motor data from the Canadian TTCP KTA 4-14 program. A schematic representation of the probabilistic service life prediction analysis is illustrated in Fig. **19**.

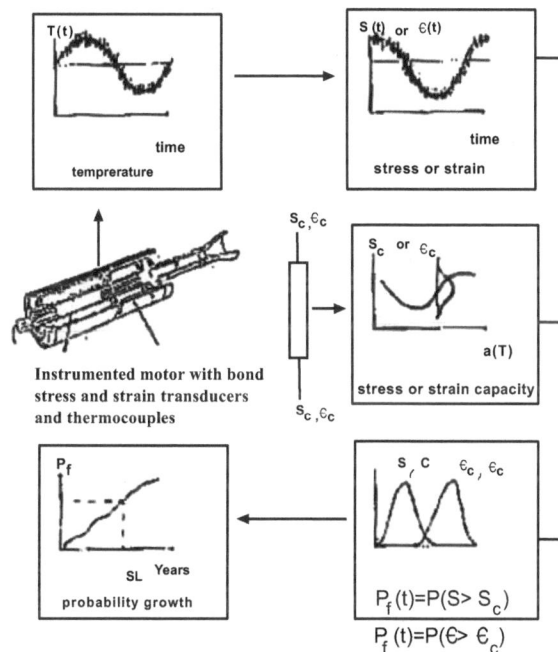

Figure 19: Schematic representation of service life prediction analysis using data from instrumented rocket motors.

Accurate service life prediction by probabilistic methods requires the following information: (1) the structural loads induced by the environment and (2) the variability of the material properties (strength capability) of the propellant. In previous studies on probabilistic service life prediction of solid rocket motors, for example reference [24], the air temperature (considered a random process) from average weather conditions is utilised in thermal stress analysis to determine the induced stresses and strains in the propellant grain. The accuracy of this method depends heavily on the accuracy of the viscoelastic models used in the thermal stress analysis. It should be noted that it is not yet possible to accurately predict the stress response theoretically for highly non-linear viscoelastic solid propellants. Also, information on the thermal environment is very scarce. Using instrumented motors would provide much more accurate data for probabilistic service life analysis as the actual and real time stresses and temperatures seen by the propellant charge are measured, alleviating the need to rely on weather conditions data to compute the thermal induced stress and strain response.

In the probabilistic service life analysis approach described in reference [4], the time-dependent randomly varying stresses and strains in the rocket motor induced by the random temperature variations is expressed through the single equation:

$$s(t) = s_m + s_y \sin\left(\frac{2\pi}{w}(t - t_0)\right) + s_d \tag{2}$$

where s_m, s_y and s_d are the stress and strain mean, yearly amplitude and diurnal amplitude respectively, t is time measured in days, t_0 is the start of the yearly temperature fluctuations and w (= 365 days) is the period of the seasonal cycle.

Assuming that the stress and strain amplitudes (s_y, and s_d) vary statistically according to the Weibull distribution, the probability density functions for s_y and s_d are given by [4]:

$$f_y(s_y) = \left(\frac{\alpha}{s_{\sigma y}}\right)\left(\frac{s_y}{s_{\sigma y}}\right)^{\alpha-1} \exp\left\{-\left(\frac{s_y}{s_{\sigma y}}\right)^{\alpha}\right\}$$

$$f_d(s_d) = \left(\frac{\beta}{s_{od}}\right)\left(\frac{s_d}{s_{od}}\right)^{\beta-1} \exp\left\{-\left(\frac{s_d}{s_{od}}\right)^{\beta}\right\}$$

(3)

where α and β are the Weibull modulus values and $s_{\sigma y}$ and s_{od} are the nornmalising factors (same dimensions as s_y and s_d).

The cumulative probability density functions, after simplification, are given by [4]:

$$P_y(s_y) = 1 - \exp\left\{-\left(\frac{s_y}{s_{\sigma y}}\right)^{\alpha}\right\}$$

$$P_d(s_d) = 1 - \exp\left\{-\left(\frac{s_d}{s_{od}}\right)^{\beta}\right\}$$

(4)

For service life prediction analysis, the propellant strength (e.g. tensile strength properties, strain capacity) is also assumed to vary statistically and has a probability density function, $f_s(S)$. This time and temperature dependent strength is gradually reduced under the influence of damage and or ageing as shown in Fig. **20**. Evaluation of this function is described in details in reference [4]. The time-dependent probability of failure at any instant of time is the probability that the induced stress (strain) response will exceed the strength (strain) capacity of the propellant, i.e.

$$P_f^i = P(S < s)$$

(5)

The instantaneous probability of failure is expressed by a triple integral relationship which can be simplified to [4]:

$$P_f^i = \int_{-\infty}^{\infty} P_S\left(s_m + s_y b(t) + s_d\right) f_d(s_d) ds_d$$

(6)

if it is assumed that the yearly amplitude s_y is deterministic and the only source of statistical variation in the stress and strain components are associated with the diurnal cycle s_d. This integral expression can then be evaluated by numerical quadrature algorithms.

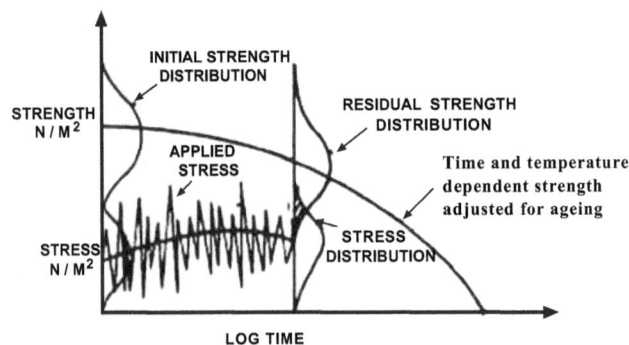

Figure 20: Applied stress and residual strength distributions.

Reference [4] developed an analysis approach which takes into account the survival of the propellant when subjected to a previous loading history. This approach used a hazard analysis to yield an expression for the failure probability after n diurnal load applications:

$$P_f\left(t_n\right) = 1 - \exp\left\{-\sum_{i-1}^{n}\frac{P_f^i}{R^{i-1}}\right\} \tag{7}$$

and $R(t_n)$, the reliability function given by:

$$R\left(t_n\right) = \exp\left\{-\sum_{i-1}^{n}\frac{P_f^i}{R^{i-1}}\right\} \tag{8}$$

The service life is the time taken for the time dependent probabilistic failure probability to exceed a specified maximum probability level P_f^{max}.

Recently Wong et al [25] has implemented new and more robust methods to compute the probability of failure. Their previous effort [4] employed Direct Numerical Integration algorithms to compute the area under the convolution integral of the joint probability density functions of the basic random variables (equation 6). This method lacks robustness and restricts the basic random variables to be described by Weibull distribution. Reference [25] describes the use of Monte-Carlo Simulation and the Advanced First-Order Reliability Method for performing the time-dependent reliability analysis and instantaneous failure probability for service life prediction of rocket motors.

5. OTHER NDE METHODS

The advantage of using NDE methods for service life assessment and/or extension of solid rocket motors is that the actual condition of the motors as it is subjected to in-service storage and operational loads is assessed. Also, the use of NDE testing in a service life surveillance program avoids the need to reduce the number of operational motors and destroy valuable assets, resulting in considerable cost savings.

Traditionally, the practice in the rocket motor community is to set up a surveillance program when life extension is required. A typical surveillance program involves withdrawing motors from deployment for static test firing followed by mechanical and chemical properties determination to validate the remaining stocks. The motors are withdrawn at regular intervals for testing over a period of 10-20+ years. A large number of motors are needed to obtain sufficient confidence in the statistical analysis.

NDE involves surveillance by using imaging techniques or non-destructive property sampling methods to assess the condition of motors aged in the field or controlled environment. Conventional NDE methods include ultrasonics and acoustic emission technique (AET) to detect defects and damage in the motor. The drawback of these techniques is that the data is often very difficult to interpret due to the low signal to noise ratio.

As mentioned in the sections above, significant advances have been made in the past two decades to instrument rocket motors with miniature stress sensors for real-time health monitoring of composite propellant rocket motors.

Other relatively recent NDE methods that have been investigated for structural integrity and service life assessment of solid rocket motors include:

- The Penetrometer [27], a self-contained device consisting of a strain-gauged indenter and microprocessor, was developed at British Aerospace Defence to monitor the mechanical properties and hence, aging characteristics of solid propellant rocket systems.

- Thermography to detect bond defects [28]. This is a very sensitive technique and works on the principle that flaws or bond defects cause resistance to heat flow and under proper conditions this causes temperature gradients to appear on the surface of the object. The temperature distribution at the surface of the object is determined by measurement of the IR radiation.

- Reference [29] demonstrated that an ultrasonic tomography technique could detect anomalies (voids and other defects) in a solid propellant from the time of arrival data by applying a pulse compression technique. However, it was noted that the resolution of the image obtained by tomographic reconstruction was limited by the angular coverage that was possible.

- Reference [30] used laser ultrasonics to determine the elastic constants and hence, damage in solid propellants. This non-contact method calculates the elastic constants from group velocity data in non-principal directions. This technique enabled mechanical properties and damage evolution to be measured simultaneously.

In the past two decades, there have been major developments in MEMs (microelectromechanical system) devices for sensors and actuators applications. It is likely that future rocket systems will be fitted with thermal and vibration MEMs sensors to record the full environmental history seen by the rocket motor. Although MEMs devices have been developed to monitor the NO_x (nitrogen oxide) evolution in the older double base propellant rockets for safe life assessment, there are no reports (to the best of the author's knowledge) on using MEMs to monitor the mechanical properties of solid composite propellant rocket motors. Some relatively recent studies on MEMs technology that may be useful to the structural integrity / service life assessment of solid composite propellant rocket motors include:

- MEMs shear stress and strain sensors have been developed and installed as arrays in a flexible skin on the surface of aerial vehicles [31].

- Wireless embedded strain sensors [32] that have been used to measure in-plane strains in laminate fibre-reinforced polymer composites.

In the future, it is envisaged that MEMs temperature and mechanical sensors could be adapted for instrumentation in solid rocket motors to monitor the temperature, vibration and induced stress / strain levels and send a signal to the monitoring station when the parameters exceed the safe or failure limits.

6. CONCLUSIONS

Rocket motor instrumentation techniques, using the miniature normal stress sensors with improved accuracy and stability, have been developed for the Pictor motor. These sensors were found to be suitable for monitoring the stresses in the motor during its service life. They were stable in the temperature range used in our environmental testing program and gave consistent data during propellant cure and environmental testing of the motor.

Results from the instrumented rocket motors for the various thermal tests can be summarized as follows: Accelerated ageing results did not show much change and bondline stresses were close to zero because the propellant is near its stress free state (i.e. the cure temperature). For the ambient ageing test, the measured bondline stresses were 50-70 kPa and did not show much change during the 280 days of storage. The thermally shocked motors showed bondline stresses in the range 350 to 400 kPa for the cooling cycle and ca. 300 kPa for the heating cycle. The stresses decreased with each thermal shock cycle due to mechanical damage. For the thermal cycling test, there was no evidence of mechanical damage in the instrumented and uninstrumented motors after 16 complete cycles.

The bond stress transducer technology provided a means of obtaining stress data directly which can be used for validating finite element stress analysis results for solid rocket motors subjected to long term thermal storage environments and also for input into probabilistic service life prediction models.

The advancement in bond stress sensor technology was further used to investigate the failure analysis of rocket motors during ignition pressurization. This is a challenging problem and the solution would lead to increased user confidence in the high strain-rate motor loading regime and more accurate service life assessments, resulting in considerable cost savings.

For this study, three motors were fitted with three normal bond stress sensors per motor in the high stress regions, using our robust installation procedure developed for bond stress sensor installation in an end burning motor. Data obtained from the static firings demonstrated the ability of the bond sensors to accurately measure the pressure in different regions of the burning motor.

Experimental stress sensor data from this study and other TTCP WTP-4 KTA 4-14 and KTA 4-23 countries with different motor configurations showed that these devices could be used in real-time motor health monitoring to detect cracks and debonds and to continuously monitor the extent of damage in the motor. These studies have also demonstrated that these miniature bond stress sensors are safe for static firing and the ability of the sensors to accurately measure pressure in different regions of the burning motor. Since these TTCP WTP-4 KTA 4-14 and KTA 4-23 studies from 1988 to 2002, Micron Instruments have made further improvements in sensor design and fabrication to eliminate thermal hysteresis and improve long term stability, to increase the accuracy and repeatability of the measurements [33].

The results from the instrumented Pictor motor verified and validated the thermal, stress and strain distributions and the regions of high propensity for crack propagation in the motor predicted by a nonlinear viscoelastic finite element analysis and modified fracture mechanics approach.

7. ACKNOWLEDGEMENTS

The author wish to acknowledge and thank Dr Frank Wong, Dr Jim Buswell, Mr Herb Chelner and Mr Peter Macdowell for all their help, support and friendship during the TTCP WTP-4 KTA 4-14 and KTA 4-23 collaborative programs. The author also wish to acknowledge and thank Mr Peter Macdowell for installing the bond stress sensors and fabrication of the instrumented motors. The author is indebted to the late Mr Gene Francis for his help and sharing of his vast experience and knowledge of rocket motor instrumentation.

8. REFERENCES

[1]   Kelly, FN, Trout JL, Elements of Solid Rocket Service Life Prediction, AIAA/ SAE 8th Joint Propulsion Specialist Conference, New Orleans, Louisana 1972; AIAA Paper 72-1085.

[2]   "Tools Required for Meaningful Service Life Prediction", JANNAF Structural and Mechanical Behaviour Working Group Publication 1974.

[3]   Francis EC, Buswell J, Service Life Prediction and Testing of Composite Rocket Motors, in Proceedings of the 20th international Conference of ICT on Environmental Testing in the 90's., Karlesruhe, Germany 1989.

[4]   Margetson, J, Wong, FC, Service Life Prediction of Solid Rocket Motors Stored in a Random Thermal Environment, in Proceedings of AGARD PEP Symposium on Service Life of Solid Propellant Systems, Athens, Greece 1996; Paper 36.

[5]   Kiezers HLJ, Miedema JR, Structural Service Lifetime Modelling for Solid Propellant Rocket Motors, in Proceedings of AGARD PEP Symposium on Service Life of Solid Propellant Systems, Athens, Greece 1996; Paper 32.

[6]   Duerr TH, Marsh BP, Sold Propellant Grain Structural Behaviour and Service Life Prediction, Chapter 5 in "Tactical Missile Propulsion", Edited by Jensen GE, Netzer DW, Progress in Astronautics and Aeronautics 1996; 170: 115-135.

[7]   Ho, SY, Francis, EC, Instrumented Rocket Motor Service Life Program, Proceedings of the 36th AIAA/ASME/SME/ASCE/AHS/ASC Structures, Structural Dynamics, and materials Conference, New Orleans, LA, USA, April (1995).

[8]   Ho SY, Carè G. Modified Fracture Mechanics Approach in Structural Analysis of Solid Rocket Motors. Journal of Propulsion and Power 1998; 14(4):409-415.

[9]   Ho SY, Macdowell P, New Service Life Methodologies for Solid Propellant Rocket Motors. DSTO Research Report, Aeronautical and Maritime Research Laboratory, DSTO-RR-0099, 1997; AR-010-186.

[10]   Ho SY, Ide K, Macdowell P, Instrumented Service Life Program for the Pictor Rocket Motor, in Proceedings of AGARD PEP Symposium on Service Life of Solid Propellant Systems, Athens, Greece 1996; Paper 28.

[11]   Ho SY, Macdowell P, Stress Analysis of Instrumented Rocket Motors Under Ignition Pressurisation Conditions. DSTO Research Report, Aeronautical and Maritime Research Laboratory, DSTO-RR-0193, 2000; AR-011-652.

[13]   Micron Instruments website http://www.microninstruments.com/store/custom-sensors.aspx

[14]   Francis EC, Briggs WE, Solid Propellant Stress Transducer Evaluation, in Proceedings of the 24th International Instrumentation Symposium, Instrumental Society of America, Albuquerque, N.M, 1978.

[15]   Marsh BP, Rocket Motor Service Life Prediction Methodologies, Vol. I, TTCP WTP4 KTA4-14 Final Report, Nov. 1994.

[16] Wong FC, Ho SY, Buswell HJ, Fillerup J, Pritchard R, Failure Analysis of Rocket Motors on Pressurisation, Vol. I, TTCP WTP-4 KTA 4-23 Final Report, Aug. 2000.

[17] Wong FC, Service Life Prediction Methodologies, Vol. III-V, TTCP WTP4 KTA4-14 Canadian Focus Officer's Final Report, Nov. 1994.

[18] Ho SY, Service Life Prediction Methodologies, Vol. II, Annex C, TTCP WTP4 KTA4-14 Australian Focus Officer's Final Report, Nov. 1994.

[19] Buswell HJ, Service Life Prediction Methodologie,s, Vol. VI-VII, TTCP WTP4 KTA4-14 UK Focus Officer's Final Report, Nov. 1994.

[20] Francis EC, Thompson RE, Bond Stress Transducer Design for Solid Propellant Rocket Motors, J. Spacecraft and Rockets, 1981; 8(5): 411-417.

[21] Wong FC, Firmin A, Liu YC, Verification of Swanson Nonlinear Thermo-Viscoelastic Model Using Stress Gauge Technology, in Proceedings of AGARD PEP Symposium on Service Life of Solid Propellant Systems, Athens, Greece 1996; Paper 26.

[22] Francis EC, Buswell HJ, Improvements in Rocket Motor Service Life Prediction, in Proceedings of AGARD PEP Symposium on Service Life of Solid Propellant Systems, Athens, Greece 1996; Paper 27.

[23] Chelner H, Buswell HJ, Miniature Sensor for Measuring Solid Grain Rocket Motor Case Bond Stress, in Proceedings of AGARD PEP Symposium on Service Life of Solid Propellant Systems, Athens, Greece 1996; Paper 26.

[24] Heller RA, Thangjitham S, Janajreh IM, Probabilistic Service Life Prediction for Solid Propellant Motors Subjected to Environmental Thermal Loads, in Proceedings of AGARD PEP Symposium on Service Life of Solid Propellant Systems, Athens, Greece 1996; Paper 34.

[25] Wong FC, Life Consumption Monitoring and Probabilistic Analysis of Munitions, Defense Research Establishment Valcartier (DREV) report 2010-01-22.

[26] Wong FC, Prediction and Measurement of Mechanical Aging in Solid Rocket Motors, Defense Research Canada Valcartier DRDC Valcartier Technical Report 2006; DRDC Valcartier TR 2010-01-28.

[27] Faulkner GS, Thompson AW, Buswell HJ, The Penetrometer: Non-Destructive Testing of Composite Propellant Rocket Motor Grains To Determine Ageing Characteristics, in Proceedings of AGARD PEP Symposium on Service Life of Solid Propellant Systems, Athens, Greece 1996; Paper 12.

[28] Schneider H, Eisenreich N, Thermographic Detection of Bond Defects Within Models of Solid Propellant Motors, in Proceedings of AGARD PEP Symposium on Service Life of Solid Propellant Systems, Athens, Greece 1996; Paper 13.

[29] Gan TH, Hutchins DA, Billson DR, Wong FC, Ultrasonic Tomographic Imaging of an Encased Highly-Attenuating Solid Media, Research in Nondestructive Evaluation, 2001; 13:3: 131-152.

[30] Bescond C, Moreau A, Lévesque D, Wong F, Bertrand L, Laser Ultrasonic Determination of the Elastic Constants of Damaged Propellant, in Proceedings of the 28th Review of Progress in Quantitative Non-Destructive Testing, Brunswick, Maine, 2001.

[31] Jiang F, Xu Y C, Weng T, Han Z, Tai YC, Flexible shear stress sensor skin for aerodynamic applications, in Proceedings of the 13th Annual International Conference on Micro Electro Mechanical Systems, 2000, IEEE, 364-369.

[32] Hautamaki C, Zurn S, Mantell SC, Polla DL, Experimental Evaluation of MEMs Strain Sensors Embedded in Composites, JMEMS, 1999; 8(3): 272-279.

[33] Chelner H, Dual Bond Stress / Temperature Sensor Performance Testing, in Proceedings of the 25th TTCP Subgroup W, Technical Panel-4 Workshop, Australia, 2000.

[34] Wong FC, Defense Research Canada Valcartier, Private Communications, 2010.

[35] Chelner H, Micron Instruments, Private Communications, 2009.

APPENDIX 1

Mıcron INSTRUMENTS

Pressure Transducer
Calibration Data

Date   18-Mar-96

Model Number	140683	Serial Number	Diaphragm Materials
Excitation	5 ma	**19810**	Steel
Pressure Range	2500 Psig		

AMBIENT PRESSURE CALIBRATION DATA   Date Calibration Run   16-Mar-96

Pressure	Increase (1)	Decrease	Increase (2)	Idea	linearly (%FS)	Hysteresis (%FS)	Repeatabrly(%FS)
0 PSIG	0.32 mv	0.22 mv		0.32 mv		0.06%	
500 PSIG	35.34 mv	35.15 mv	35.24 mv	34.704 mv	0.37%	0.11%	0.06%
1000 PSIG	69.91 mv	69.67 mv	70 mv	69.088 mv	0.48%	0.14%	0.05%
1500 PSIG	104.22 mv	104.06 mv	104.22 mv	103.472 mv	0.44%	0.09%	0.00%
2000 PSIG	138.37 mv	138.31 mv	138.39 mv	137.856 mv	0.30%	0.03%	0.01%
2500 PSIG	172.24 mv		172.3 mv	172.24 mv			0.03%
Sensitivity	171.92						

STATIC ERROR BAND ±   0.24% B.F.S.L.FS

	30 deg F	80 deg F	130 deg F
0 PSIG	0.89	0.77	-0.02
250C PSG	171.81	172.2	-78.6
Sensitivty	170.92	171.43	170.52

Thermal Balance Shift= .0027% deg F
Thermal Sensitivity Shift= .0018% deg F

Mıcron INSTRUMENTS

Pressure Transducer
Calibration Data

Date   18-Mar-96

Model Number	140683	Serial Number	Diaphragm Materials
Excitation	5 ma	**19811**	Steel
Pressure Range	25C0 Psig		

AMBIENT PRESSURE CALIBRATION DATA   Date Calibration Run   16-Mar-96

Pressure	Increase (1)	Decrease	Increase (2)	Idea	Linearity(%)	Hysteresis (%FS)	Repeatabrly(%FS)
0 PSIG	0.35 mv	0.22 mv		0.35 mv		0.08%	
500 PSIG	30.55 mv	30.4 mv	30.51 mv	31.146 mv	0.39%	0.10%	0.05%
1000 PSIG	61.32 mv	61.2 mv	61.26 mv	61.942 mv	0.40%	0.08%	0.04%
1500 PSIG	92.51 mv	92.32 mv	92.44 mv	92.736 mv	0.15%	0.12%	0.03%
2000 PSIG	123.61 mv	123.36 mv	123.55 mv	123.534 mv	0.05%	0.16%	0.02%
2500 PSIG	154.33 mv		154.32 mv	154.33 mv			0.01%
Sensitivity	153.98 mv						

STATIC ERROR BAND ±   0.20% B.F.S.L.FS

	30 deg F	80 deg F	130 deg F
0 PSIG	0.14	0.39	0.28
2500 PSIG	153.08	153.98	153.56
Sensitivity	152.94	153.59	153.28

Thermal Balance Shift= .0006% deg F
Thermal Sensitivity Shift= .0021% deg F

CHAPTER 7

Solid-Fueled Rocket Structural Integrity Assessment

S.Y. Ho

Defence Science and Technology Organisation, P.O. Box 1500, Edinburgh, SA 5111, Australia.

Abstract: This chapter gives an overview of the main uses and features of structural integrity assessment (SIA) of solid rockets. Details of the key elements and prediction techniques for analyzing the stress and strain response, failure criteria and the service environments in which solid rockets are required to operate are given. The development and use of these methods are essential to deal with the increasingly more stringent requirements in performance, safety, reliability and cost. The role and extent to which SIA is employed in design and development, service life prediction and vulnerability / safety assessment of solid rockets are discussed. The development of appropriate nonlinear constitutive models (temperature and strain-rate dependent and incorporates damage) and their implementation in finite element codes, suitable failure criteria (e.g. based on facture mechanics), and service life prediction methodologies are discussed, and an example solution procedure is provided for a real-life rocket. A method for determining high strain-rate mechanical properties, using the Hopkinson Bar technique, to assess the structural response of rocket component materials to impact loading conditions is described.

1. INTRODUCTION

The structural integrity of the propellant grain governs the performance, reliability and service life of the solid rocket propulsion system. This chapter describes the design features, service life and vulnerability assessment of the solid rocket in the context of structural integrity and the service environments in which the solid rocket is required to operate. Compared to liquid rockets and air-breathing propulsion systems, solid rockets are relatively simple in design (e.g. see Fig. 1). They provide the easiest means to achieve the high mass flow rates and thrusts required for the initial boost phase of launch vehicles, missiles and air-breathing sustainers. Hence, it is expected that the solid rocket will continue to have a prevalent role in the future of the aerospace and defence industry.

Important functional considerations in the design of solid rockets must not only include propulsion performances but also structural integrity and operational requirements (such as environmental conditions, signature of combustion products, safety / vulnerability and reliability). With the introduction of Insensitive Munitions (IM) requirements [1] in many countries, there is a growing need to assess the probability and consequence of failures for safety reasons. Because of their increasing importance, the vulnerability requirements (which are part of the operational requirements) will be discussed in more detail in Section 3. Only those vulnerability issues related to structural integrity / failure of the solid rocket are considered in this chapter.

Solid rockets are required to operate in more severe and wide-ranging environments than the other forms of propulsion systems. During the service life of a solid rocket, the propellant is subjected to stresses which can cause cracks in the grain or separation between the propellant and the inhibitor or the insulation. There are a number of consequences of these structural failures during rocket firing. Damage in the propellant grain generally leads to excessive combustion pressure and violent rupture of the motor case, resulting in a mission failure or, in the worst case, a catastrophic failure with loss of platform and lives. Therefore, it is necessary to design and verify for high levels of reliability for solid propellant grains.

The structural integrity of the solid rocket may be significantly affected by environmental conditions and this must be taken into account in the design as well as in the service life estimate (i.e. length of time for the system to reach safe life limit or a critical design condition) required by the customer. In the design phase, structural integrity assessment (SIA) of the propellant grain is performed to evaluate the capability of the solid rocket motor to meet the mission requirements under the specified operational and storage environments. In the service life assessment (SLA) of solid rockets, it is also important to consider ageing of the propellant in the SIA, which results in degradation of mechanical and /or chemical properties with time. In the SLA, the structural integrity of the ageing propellant grain under the service environmental loads is verified and the length of time for the propulsion system to reach the safe life limit determined.

This chapter describes the main uses, key elements and analyses / prediction techniques for structural integrity assessments of solid rockets. The state-of-the-art and challenges in determining /developing constitutive and failure models for the non-linear viscoelastic motor components under appropriate loading conditions, structural integrity and service life assessment of solid rockets are presented.

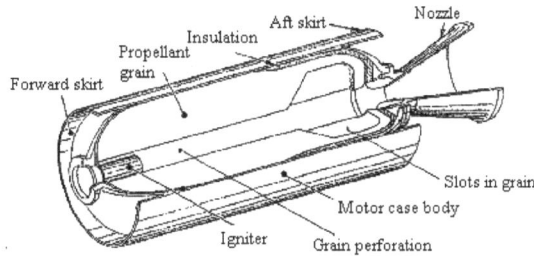

Figure 1: Example of a solid rocket motor with slots in propellant grain.

2. STRUCTURAL INTEGRITY ASSESSMENT AND SERVICE LIFE PREDICTION

In solid rockets (whether the propellant is of the double-base or composite type and whether the grain is cartridge loaded or case-bonded) SIA is used in the various stages in the research, development and use of the grain in a particular application. It is utilised during the design and development processes to provide a systematic analysis of the possible failure modes under the required loading conditions to determine the risks in the motor development and service use. Hence, it enables the required extent of testing to be determined for verification of the rocket motor's structural integrity before entry into service, as well as a more accurate estimate of the service life. It also provides a useful basis for structural failure investigations.

The two main aspects in the structural integrity assessment are: (1) structural analysis which determines the stress, strains and deformations in the solid rocket grain under the specified loading conditions during its life cycle and (2) failure analysis which determines the strength capability of the various structural components in the rocket (e.g. propellant, bondline materials, rocket casing). The predicted stresses or strains from the structural analysis are then compared with appropriate failure criteria from the failure analysis to evaluate a grain safety margin or a minimum safety factor for rocket motor operation. Fig. **2** shows a general outline of a structural integrity assessment procedure and some typical data required to conduct the analysis.

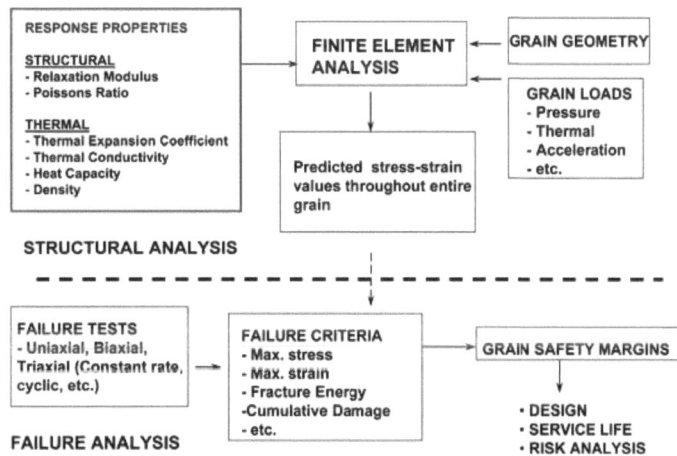

Figure 2: Outline of a typical structural integrity assessment procedure for solid rocket motors.

2.1. Structural Analysis

There are a number of different approaches for structural analysis of the solid rocket propellant grain. The procedures generally utilised by the solid rocket motor industry / community are described below. The structural analysis of propellant grains is unique, primarily due to its nonlinear viscoelastic material behaviour [3-10]. With recent dramatic improvements in computational capabilities and the development of more accurate constitutive laws, finite element based analytical methods are now extensively used for structural analysis of solid rockets.

2.1.1. Finite Element Model Definition

Only a brief discussion on the finite element method (FEM) pertinent to the structural analysis of solid rockets is given here, as excellent resources describing the method in much greater detail can be found elsewhere [11-13].

To successfully compute a solution, the FE model needs to include properly defined geometry, material properties and boundary conditions, such as applied loads and constraints to prevent rigid body motion. The problem can be simplified if the grain geometry has planes of symmetry. If the model geometry, material properties and boundary conditions do not vary longitudinally the geometry of the FE model can be restricted to two dimensions. Furthermore, the assumptions of (1) plane stress / strain or (2) axisymmetric geometry and loading can be used to restrict the problem to a 2-D FE analysis. In some cases (e.g. when grain geometry has non-axisymmetric features, non-uniform loading conditions) it is necessary to perform a full 3-D FE analysis. After the FE model is defined, critical regions (e.g. bondlines) in the model should be identified and the finite element mesh refined in these areas for higher accuracy.

Typical 2-D and 3-D elements used for modeling solid structures are shown in Fig. **3**. Element selection affects the accuracy, speed and numerical stability of the solution. It requires consideration of shape factor, integration rule and so on, since some combinations of elements may be computationally inefficient or produce numerical instabilities.

ELEMENT	TYPES	Example
2D Plates • nodes x integration points	3 and 6-node triangle	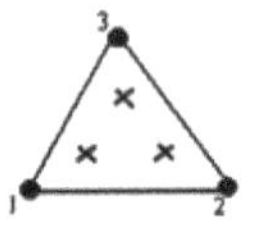
	4, 8 and 9-node quadrilateral	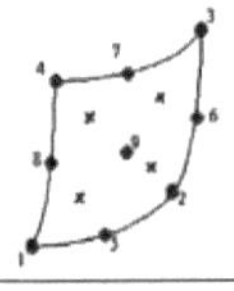
3D Bricks	4 and 10-node tetrahedral	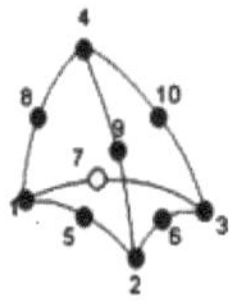
	6 and 15-node wedge	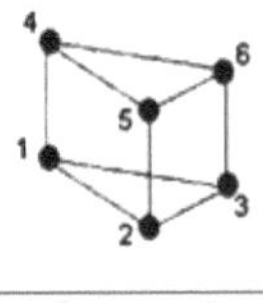
	8, 16 and 20-node hexahedral	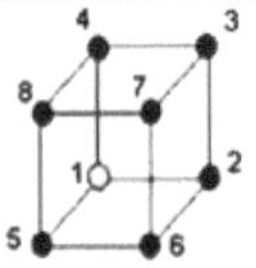

Figure 3: Typical two- and three-dimensional element types for structural analysis of a solid rocket motor.

2.1.2. Response Properties and Material Definition

A major challenge in the structural analysis of solid propellant grains is modeling the complex material behaviour under various loading conditions. Modern solid propellants are composite materials containing 10-20% by weight of a rubber-based material (e.g. hydroxyl terminated polybutadiene) which acts as a binder for the remaining large percentage of solid oxidizer (e.g. ammonium perchlorate). Under certain loading conditions, the interaction between the oxidizer particles and the viscoelastic polymeric binder can result in dewetting (debonding of the binder from the oxidizer particles) and void formation. Hence, solid propellants exhibit nonlinear viscoelastic behaviour. The four major sources of solid propellant nonlinearities are: 1) modulus strain sensitivity, (2) volume change / dewetting, (3) thermo-mechanical coupling and (4) damage and rehealing. The choice of a material constitutive law depends on the loading condition and loading history. It is beyond the scope of this chapter to cover all the nonlinear theories which have been developed for solid propellants. Some of these nonlinear constitutive theories are:

Hyperelastic models coupled with viscoelasticity, e.g. the Mooney [14], Rivlin [15], Odgen [16] and Neo-Hookean models have been used to describe nonlinear elastic behaviour.

The Swanson and Christensen model [17] used the Green-Lagrange definition of strain which differs from the small strain relationships in that the second order derivatives of displacement are retained, permitting the rotational geometric nonlinearities to be taken into account. The Simo model [19] uses decoupled volumetric and deviatoric responses. The Farris model uses time-independent bulk response and time-dependent deviatoric response [3-5]. The Lee model [6] accounts for coupled thermomechanical effects.

Relatively recent methods for modeling the nonlinear viscoelastic behaviour of solid propellants include an improved Mori-Tanaka model proposed by Wong [9-10] and a unified model that incorporates damage and nonlinear viscoelastic response proposed by Ho [8]. This unified constitutive model was developed for implementation in a finite element code and will be described in more detail in the section below.

2.1.2.1. Constitutive Laws

Linear Viscoelastic Model

Solid rocket motor components, such as the propellant, inhibitor, insulation and propellant-inhibitor bondline, exhibit thermoviscoelastic material behaviour, i.e. their response and stress-strain capability are time and temperature dependent [5, 20, 21]. Linear viscoelastic constitutive models are commonly used to describe the stress-strain relations of solid rocket propellants, and are valid at low levels of strain, in the linear region before dewetting and other nonlinear effects occur. For linear viscoelastic materials, the stress is related to an arbitrary strain history through the convolution integral:

$$\sigma = \int_0^{\xi} E\left(\xi - \bar{\xi}\right) \frac{d\varepsilon}{d\bar{\xi}} d\bar{\xi}$$ (1)

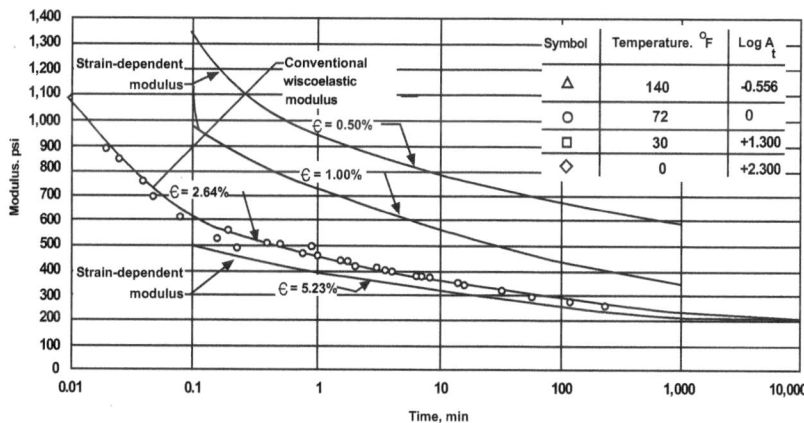

Figure 4: Conventional viscoelastic modulus data and strain dependent modulus for solid propellants. Courtesy of E.C. Francis [3].

where $\bar{\xi} = t/a_T$ is the reduced time and $E(\xi)$ is the relaxation modulus obtained from stress relaxation tests. It is assumed that the temperature dependence can be separated from the time dependence through the use of a time-temperature shift factor, a_T, and the real time, t, can be substituted by a reduced time $\bar{\xi}$.

Fig. **4** shows some data for conventional viscoelastic modulus and strain dependent modulus of solid propellants.

Stress relaxation tests, dynamic mechanical thermal analysis (DMTA) or other mechanical properties measurements over a wide range of temperatures and strain-rates must be performed to experimentally determine the propellant response properties from master modulus curves (for example, see Figs. **5** and **6**). Time-temperature superposition (based on the assumption that a shift in temperature is equivalent to some shift in time) is applied to materials where this principle is viable, i.e. their response and failure behaviour show a dependence on temperature and time (rate).

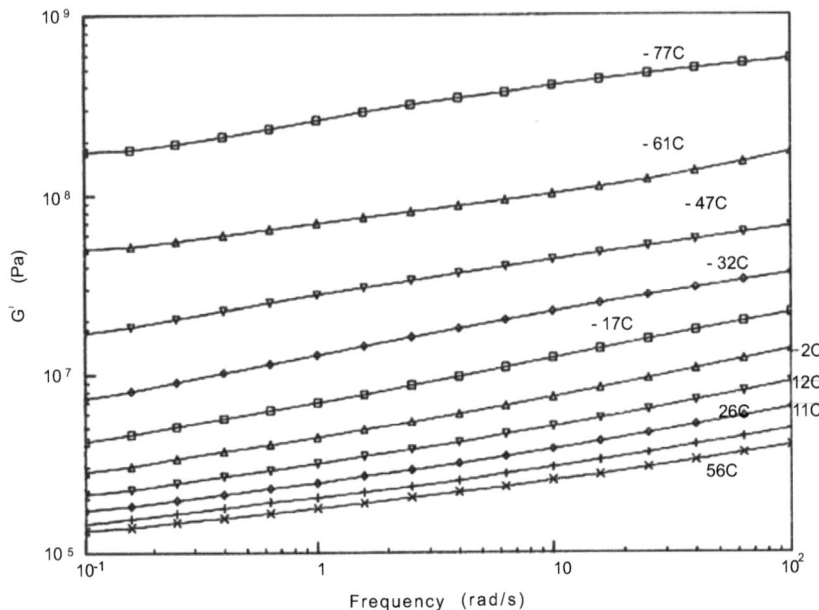

Figure 5: Dynamic shear modulus as a function of frequency at various temperatures, as taken from reference [23].

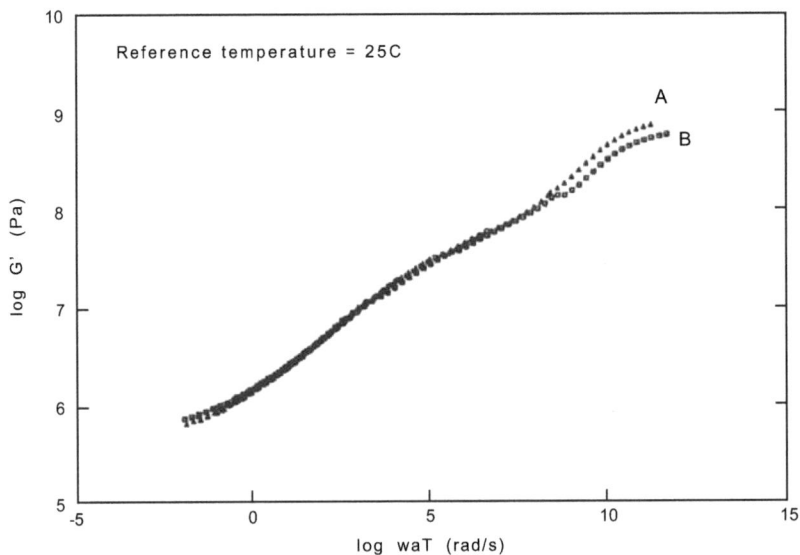

Figure 6: Master curves of log (dynamic shear modulus) vs. log (frequency * a_T) for two composite propellants (A and B), as taken from reference [23].

In Fig. **6**, master curves were obtained by horizontally superposing the modulus data at different temperatures onto a composite curve at a reference temperature of 25°C [21]. The constants C_1 and C_2 in the WLF equation (a frequently used representation of the shift factor a_T as a function of temperature) were determined by fitting the horizontal shift factors to a transformed form of the equation, viz:

$$(T - T_0)\log a_T = -C_1(T - T_0) - C_2(\log a_T) \tag{2}$$

where T is the temperature at which the experiment was conducted, T_0 is the reference temperature and a_T is the shift factor. The best values of C_1 and C_2 were determined from a multilinear regression analysis. Similarly, a stress capability master curve can be used to determine stress capabilities under different operational and storage conditions. It should be noted that some solid propellants exhibit a thermorheological complex behaviour where the time-temperature superposition is not viable, and other constitutive models must be used to describe their stress-strain relationship.

The temperature and rate dependent modulus data from the master curve is used in the structural analysis of the propellant grain to determine the response behaviour of the propellant under various operational and storage loading conditions. Similarly, the stress capability master curve is used in the failure analysis to determine the failure properties under various loading conditions.

Combined Nonlinear Viscoelastic and Damage Model

The unified constitutive model in reference [8] incorporates mechanical damage, and temperature and strain-rate dependence. Although it was originally developed for the high strain-rate regime, it is also applicable for low strain-rate loading conditions. The features of this model are: (1) the numerical parameters of the model can be related to the physics of the propellant and obtained by multivariate regression analysis of the experimental data from tests performed under various strain-rates and temperatures. (2) Temperature and strain-rate dependence is handled using the time-temperature superposition principle. (3) Damage is incorporated in the constitutive equation using the "stress correction" approach of Swanson and Christensen [2, 17] where a stress softening function is used to correct the stress response. (4) A continuum energy-based approach is used to determine the stress softening data from simple loading experiments. A suitable function, which can account for the effect of different loading variables such as strain-rate, to fit the stress softening data is determined.

For a given temperature, T, and strain-rate, $\dot{\varepsilon}$ (s^{-1}), the equivalent stress at a given strain level can be calculated from the master curve

$$\sigma_R = E_R \varepsilon^m + \eta_R (\dot{\varepsilon}\, a_T)^n \tag{3a}$$

where E_R (E is the modulus from the linear elastic region of the stress-strain curve at a given temperature and strain-rate) is obtained from the master curve (Fig. 7), and η_R ('pseudo' or artificial velocity), m (exponent for strain softening or hardening; m < 1 for strain softening and m >1 for strain hardening) and n (exponent for strain-rate dependence) are the fitted coefficients from the stress-strain curve at the reference temperature. The shift factor can be calculated from equation (2) using the WLF constants. The constitutive equation (3a) retains its predictive capability, and is suitable for computational purposes to implement in finite element codes for predicting the structural response of rocket motors [7].

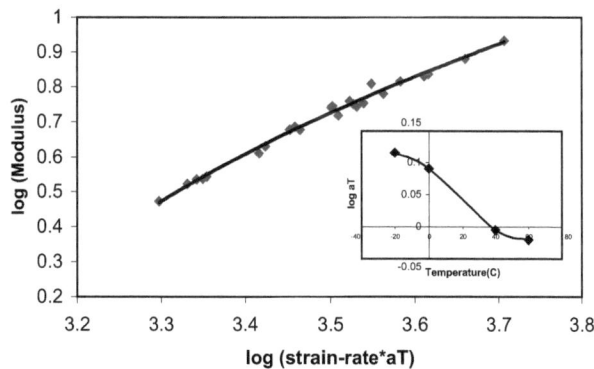

Figure 7: Master modulus curve and shift factors (inset) vs. temperature for a typical solid composite propellant, as taken from reference [8].

Inclusion of a softening function in the constitutive equation (3a) allows damage to be taken into account, as well as temperature and strain-rate dependence [8].

$$\sigma_D = g(E_R \varepsilon^m + \eta_R (\dot{\varepsilon}\, a_T)^n) \tag{3b}$$

Here, σ_D is the equivalent stress of the damaged material at the reference temperature. The softening function, g, is defined by Swanson and co-workers [17] as the ratio of the observed stress to the viscoelastic stress and is equal to unity for an undamaged propellant.

In reference [8] an energy approach, based on recoverable strain energy density [20, 22] was used to determine the softening function (g) for damaged propellants (induced by constant stress tests [18]) from a calibration model, such as equation (4).

$$g = AW_{rc} + B\dot{\varepsilon} + C \tag{4}$$

where W_{rc}, the recoverable strain energy density, and $\dot{\varepsilon}$ are the independent variables of the multivariate function. It is believed that an energy-based approach can better capture all the damage in the propellant because it takes into account both the stress and strain capability and is applicable to different loading conditions. Another advantage of this approach is it allows for bulk inelastic behaviour (ie. energy dissipated by plastic deformation, etc.).

The recoverable strain energy density is given by:

$$W_{rc} = W_i(1 - H_r) \tag{5}$$

where W_i is the input strain energy density and includes the stored energy available for crack growth as well as energy dissipated in viscoelastic processes and plastic deformation. The hysteresis ratio, H_r = *(loading energy - unloading energy)/ loading energy*, is a measure of the non-recoverable energy losses obtained from load-unload experiments [22].

The input strain energy density, $W_i = (\int \sigma d\varepsilon)$, was calculated from the stress-strain curve of the propellant. The hysteresis ratio of the solid propellant must be characterized in order to calculate W_{rc}. Typical hysteresis ratios for composite propellants range from 0.3 – 0.45 [22].

For the specimen geometry and loading conditions in reference [8], the function g determined using equation (4) is a piecewise function where

$$g(W_{rc}, \dot{\varepsilon}) = \begin{vmatrix} 1.0 - 6.6 \times 10^{-6} W_{rc} - 48\dot{\varepsilon}, & 0 \leq W_{rc} \leq 1.8 \times 10^3 Jm^{-3} \\ 1.7 - 4.1 \times 10^{-4} W_{rc} + 1.0 \times 10^4 \dot{\varepsilon}, & W_{rc} > 1.8 \times 10^3 Jm^{-3} \end{vmatrix} \tag{6}$$

Fig. **8** shows a plot of g vs. W_{rc} for constant stress tests conducted in the strain-rate range representative of the loading conditions that induced damage in the propellant. As indicated by equation (6) and illustrated in Fig. **8**, the softening function decreased markedly with increasing applied strain energy density, above a recoverable strain energy density level of 1.8×10^3 Jm^{-3}, suggesting two different damage mechanisms. A sufficient amount of predamage was required, that is, the applied strain energy density must exceed a certain level, before the damage was manifested as a reduced stress-strain response. The initial decrease in the slope of the g vs. W_{rc} plot was attributed to dewetting and the formation of voids and microcracks. The rapid decrease in the slope once a certain applied strain energy density level was exceeded was attributed to crack propagation and other secondary deformation mechanisms [8].

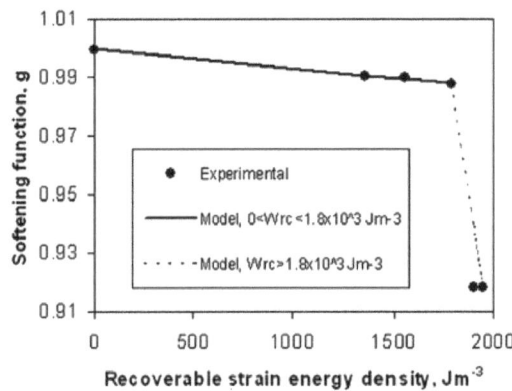

Figure 8: Stress softening function vs recoverable strain energy density W_{rc} from constant stress tests, as taken from reference [8].

The unified constitutive model, equation (3a, b), from reference [8] was shown to provide reasonably good correlation between the predicted and experimental stress response. This indicated that firstly, within the appropriate temperature and strain-rate ranges, equation (3a, b) could be used for temperature-rate conversion. Secondly, a stress softening function, characterized from simple tests, that was based on an energy approach could be used to predict damage. This method of incorporating damage in the viscoelastic functions necessitates separate tests to characterize the softening functions and the viscoelastic parameters. The viscoelastic losses are time dependent, but recoverable. On the other hand, the softening function is a measure of non-recoverable energy attributed to dewetting, formation of voids/microcracks and crack propagation. A universal model linking these different phenomena from a single test may require microstructural constitutive theory.

2.1.3. Environmental Load Definition

Verification of the structural integrity of solid rockets during design, testing, hardware development and service life assessment requires knowledge of the service environments, which includes both storage and operational loading conditions. Thermal loads from long term exposure to a wide range of temperature during field and fleet storage typically represent the most demanding loading conditions. Flight loads from motor ignition and operating pressure, and axial acceleration during launch are probably next in importance.

There are three broad classes of service environments for tactical rocket systems: land, air and sea. Nowadays, many countries are storing their land-based rocket systems in temperature controlled magazines under specified conditions. Air-based rocket systems are subjected to thermal shock conditions during air carriage and vibration loading during flight. Forced cooling in high altitude low speed flights can result in very large temperature gradients and consequently, large thermal induced stresses in the rocket motor. Ship-based rocket systems are normally stored in air-conditioned magazine and the operational loads include vibration and shock. All these service and operational environmental loads need to be considered in the structural analysis of the rocket.

The following paragraphs discuss some common practices used in the implementation of the loading conditions into the finite element structural analysis, where it is often necessary to make some assumptions because of the complexities arising from modeling combined loads, dynamic effects, and so on.

Thermal Loads

Thermal induced stresses and strains are caused by the difference in thermal expansion coefficients between the different rocket motor components. The thermal expansion of propellants and inhibitor and insulation materials are generally an order of magnitude greater than that for the motor case materials [26]. Hence, temperature gradients can produce significant bondline stresses. Critical loading conditions can arise from thermal cooldown (e.g. during propellant curing), thermal cycling and thermal shock. For finite element analysis of thermally cycled rocket motors, it is important to consider the time-temperature dependence of the behaviour of the propellant material and use a viscoelastic constitutive model for the propellant.

An important parameter for calculating thermal loads is the stress free temperature. The propellant cure temperature is usually taken as the stress free temperature. In reality, however, the stress free temperature changes with time (shifts towards the storage temperature) and may not be constant throughout the propellant grain.

Aerodynamic Heating

For air carriage tactical rocket systems, heating of the external surfaces can cause a large thermal differential and induce a significant load on the bondline of case-bonded rockets. This results in expansion of the motor case with a decrease in the strength capability of the bondline during the temperature rise. More details on the aerodynamic heating analysis can be found in chapter 3 of this book.

Acceleration and Slump (storage) Loads

The load from axial acceleration during rocket launch is usually not critical since significant grain burnback usually occurs before critical acceleration loading condition is reached. Also, the internal combustion pressure subjects the grain to a high state of compression. However, for flight loads, maximum shear stress may occur near the time of maximum pressure (generally immediately after ignition rise time for a simple single thrust or dual thrust boost-sustain motor). The shear stresses are usually highest at the bondline interfaces, particularly with acceleration at high temperatures, since the bondline strength is reduced as temperature is increased. The regions of localized stress gradients should be analysed by detailed finite element analysis.

Acceleration loads (slump) resulting from vertical or horizontal storage is typically 1.0G and the induced stress and strains are typically at a low level for small size tactical rocket motors but can be significant for large booster rockets for launch vehicles. Propellant softening from long term ageing or high temperature storage (because of the long time creep behaviour of the propellant) can cause significant slumping and it is important to consider this effect during long term horizontal storage of large motors.

Ignition Pressurisation

These loads occur during ignition and during motor burning. They impose hydrostatic compressive stresses on the burning surface of the propellant grain. This is typically modeled as a static load, taken as the initial maximum pressure obtained from static firing or interior ballistic prediction.

Vibration and Shock Loads

An analysis of the vibration loads on the solid rocket motor may be required if vibration due to air carriage, ship motion or during transportation is significant. Solid rocket motors instrumented with accelerometers can be used to measure the vibration spectrum [26]. Vibration loads generally induce the bending modes [27] and the induced stresses are small compared with thermal, acceleration and pressure loads.

Shock loads imposed by accidental dropping of rockets during transport should also be considered. This will be discussed further in Section 3 of this Chapter.

Combined Loads

An example of combined loading condition is thermal cooldown and ignition pressurization. The effects of combined loads need to be considered to accurately reflect the true loading condition. The results from the thermal and ignition pressurization analyses are obtained separately and then superposed to obtain the stresses and strains for the combined loading condition (assuming the Boltzmann superposition principle is valid for the material). Alternatively, a transient analysis can be carried out on a thermal loading followed by a pressure loading at a specified time using user-defined algorithms in the FE code [7, 43]. In the case of nonlinear materials, an appropriate constitutive law must be chosen which can handle multiple loads.

2.1.4. Example Solution Procedure

Model definition

The following example [28] illustrates the finite element structural analysis of a relatively large generic rocket motor with a nine-point propellant grain geometry as shown in Fig. **9**. The model consisted of the cylindrical motor case, insulation layer, split-boot region and propellant. The end closures were not modeled, as only stresses and strains in the propellant region were of interest. Due to the localised nature of the regions of interest and structural symmetry, the computational mesh used in the FE analysis can be significantly reduced as shown in Fig. **10**. The nine-point grain geometry allows for a single 40 degree sector to be modeled instead of the circular whole. The cross section is essentially uniform throughout the majority of the motor length. It has only a slight taper (a few millimetres over its total length) and cross sectional variation is restrained to a split-boot region, identical at each end of the cylindrical case. This allowed for great simplification in modeling, as only the region in the vicinity of the ends need be modeled. The model is five times long as it is high, to allow for St Vennant's principle effects. This was verified in the final solution. The effects of taper in the propellant bore were neglected, as the taper was so slight.

Figure 9: Propellant cross section.

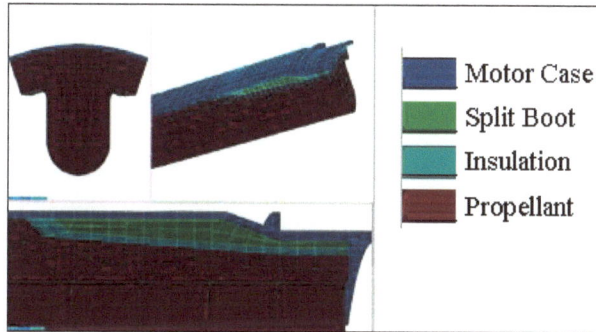

Figure 10: Finite element mesh geometry

Displacement Constraints

The nozzle end of the model was rigidly fixed at nodes corresponding to the threaded joints. The presence of the end closure in the actual motor would provide extra stiffness here and prevent any movement. The propellant at this end was a free surface, as in the real rocket. The opposite end of the model represented an effective centre of the rocket, that is, a symmetry plane. Appropriate symmetry conditions were applied. The two sides of the 40 degree wedge were assigned sector symmetry constraints.

Thermal Load Cases

The load cases applied to the model were chosen to correspond to the lower and upper temperatures in the design operational requirements. The two thermal loads used in the 3-D viscoelastic analysis were: (1) ambient to +74°C followed by (2) ambient to –54°C, representing the upward and downward thermal transients in a severe thermal shock test.

Finite Element Mesh

Due to the complex three-dimensional nature of the model, brick elements were required to adequately capture the physics of the structure. Twenty-node brick elements and fifteen-node wedge elements were utilised. Mesh refinement was carried out in the region of the split boot, see Fig. 10. This is where the geometry changed the most, and where stresses and strains were likely to build up. In all cases, an aspect ratio of 10:1 or greater was avoided with the bricks, as this introduces errors in the stiffness matrix calculation within the solver.

Material Response Properties

The response of the component materials in a rocket motor is complex, and is dependent on the load conditions (e.g. direction, magnitude), environmental conditions (e.g. temperature, pressure) and loading history. Hence, it is important to use the correct mathematical description of the material mechanical/thermal behaviour in the material constitutive laws.

The relaxation modulus vs time data from the stress relaxation experiments were used to develop propellant constitutive models. Specific heats, thermal expansion coefficients, thermal conduction coefficients and calculated convection coefficients as a function of temperature were input values used in the non-linear transient heat analysis.

Thermal-structural Analysis

A nonlinear transient heat analysis was conducted using the two thermal load cases described above, followed by nonlinear transient dynamic analysis to determine the stresses and strains induced in the motor by the thermal loads.

Figs. 11 and 12 show the temperature distribution in the 3-D finite element model during the thermal shock loading from 25°C to -54°C. The model predicts that after 2 hours the temperature difference between the propellant adjacent to the motor case and at the bore surface is around 26°C. Thermal equilibrium is reached after approximately 30 hours.

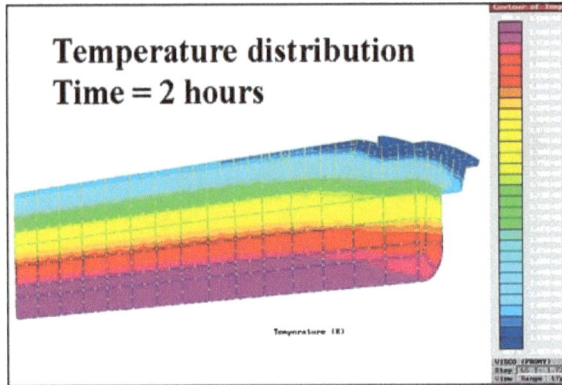

Figure 11: Temperature distribution at time = 2 hrs, from nonlinear transient heat analysis.

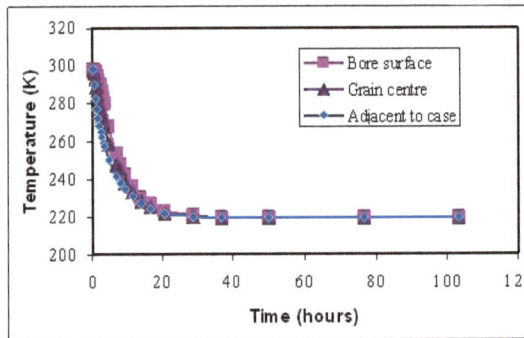

Figure 12: Temperature history from nonlinear transient heat analysis.

The von Mises stress and strain contours at the end of the downward (ambient to -54°C) thermal loading are shown in Figs. **13** and **14**. The non-linear viscoelastic solutions assume that the motor begins in a state of zero stress and strain at the start of the thermal loading.

The high stresses in the propellant grain are located adjacent to the steel casing and insulation and boot material. This is where the grain geometry changes, that is, at the boot-grain interface; and is where stress concentrations are expected. The regions of highest strain are also localised to the split boot region, which is acceptable given the rubbery nature of the material in this area. Strains in the propellant are again greater at the boot-grain interface.

As expected, the induced stresses for the hot loading are compressive and close to zero because the propellant grain is near the stress free temperature (usually a few degrees below the cure temperature). The stresses in the cold loading are tensile and are two orders of magnitude higher than in the hot loading. The cold loading case is more critical as propellant grain cracks are more likely to initiate at colder temperatures.

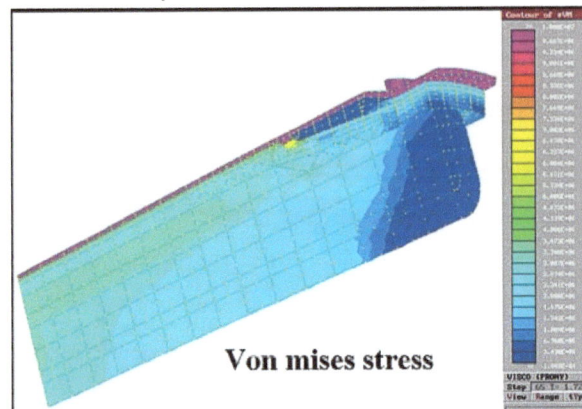

Figure 13: Von Mises stress contour at the end of a thermal loading from 25 to -54°C.

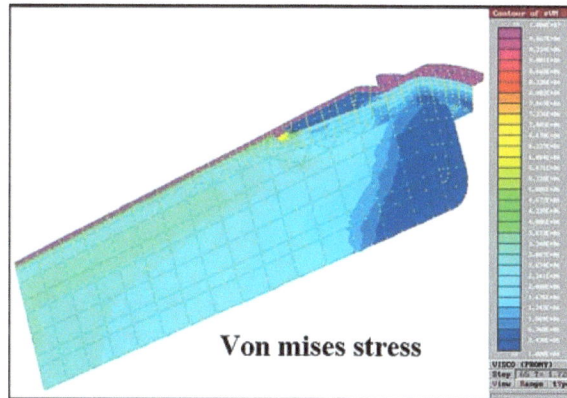

Figure 14: Von Mises strain contour at the end of a thermal loading from 25 to -54°C.

2.2. Failure Criteria

A failure criterion is a model that utilizes data from simple test conditions to predict failure under complex conditions. The material type, loading conditions, location of failure and the type of failure influence the application of failure criteria. The failure criterion applied would depend on the established failure mode. For example, in solid rockets maximum principal strain is normally used for failure in the grain bore while maximum principal stress is used to analyse cracking in the grain or bondline.

There are two basic categories of failure criteria according to the presence or lack of an initial flaw (microcracks/cracks, voids, etc.). When pre-existing flaws are not large enough to influence fracture (crack propagation), failure is dominated by the ultimate strength of the material. Most of the classical approaches used to predict failure are based on single loading or initial loading conditions. Some commonly used failure criteria for unflawed propellants are described below.

Maximum principal stress – failure is predicted when the largest principal stress exceeds the maximum principal stress from uniaxial tensile test.

Maximum deviatoric stress – failure is predicted when the maximum deviatoric stress exceeds the largest devatoric stress from pressurised uniaxial tensile test ($\sigma_1^{DEV} \geq 2/3\ \sigma_0$, where σ_1^{DEV} is the maximum deviatoric stress and σ_0 is the maximum stress from a pressurised uniaxial test).

Maximum principal strain - failure is predicted when the largest principal strain exceeds the maximum principal strain from uniaxial tensile test.

Maximum shear stress (Tresca) – failure is predicted when the maximum shear stress exceeds the maximum shear stress from a uniaxial tensile test ($\tau_{max} \geq 1/2\sigma_0$, where τ_{max} is the maximum shear stress and σ_0 is the maximum stress from a uniaxial tensile test).

Strain energy – failure is predicted when the strain energy density exceeds the measured strain energy density from a uniaxial tensile test ($U_{computed} \geq U_m(R^*a_T)$ where U_m is the measured allowable strain energy density determined for the reduced strain-rate, R^*a_T, from the master uniaxial strain energy density curve and strain energy density is defined by $U = \int \sigma d\varepsilon$ where σ and ε are the stress and strain respectively.

Smith failure envelope – failure envelope is defined by a plot of failure stress vs. failure strain. The data is obtained from constant strain-rate, constant stress rate, constant stress and constant strain tests and hence, the failure envelope is independent of time and temperature.

It is well known that all propellant grains have flaws (voids, microcracks) that result from processing and/or prior loads. If the flaws/cracks are above a critical value, fracture mechanics can be applied to determine whether the cracks will grow on subsequent loading. Damage models are used to describe failure from multiple loadings. The different types of failure criteria for flawed propellants are summarised below.

Critical stress intensity factor (K_{IC}) - only applicable to materials which exhibit linear elastic behaviour, e.g. brittle materials such as the double base type propellants. The crack will propagate if the stress intensity factor reaches a critical value given by $K_{IC} = Q\sigma_c a^{1/2}$ for mode I loading where Q is a geometry parameter, σ_c is the applied stress at the onset of crack growth and a is the crack length.

Critical strain energy release rate or fracture energy (G_c). The energy criterion for fracture is given by:

$$\partial(F-U)/\partial a \geq \gamma (\partial A/\partial a) \tag{7}$$

where ∂A is the increase in surface area associated with an increment of crack growth ∂a, F is the work done by the external force, U is the elastic stored energy in the bulk specimen and γ is the surface free energy. Assuming that the energy dissipation around the crack tip is independent of test geometry and loading conditions, 2γ can be replaced by G_c, the energy required to increase the crack by unit length. The fracture energy approach can be readily extended to inelastic materials [7, 24], where G_c is now the total amount of energy dissipated during crack growth and includes energy dissipated by plastic deformation and viscoelastic processes. Fracture energy is considered to be a more complete failure criterion than those criteria based on stress or strain capabilities alone.

J integral or crack extension force – only valid for linear and non-linear elastic materials. The crack extension force in the x_1, x_2 directions can be defined by a path independent expression

$$J = \int_\Gamma W dx_2 - \sigma_{ij} \frac{du_i}{dx_1} d\Gamma_j \tag{8}$$

where W is the strain energy density, σ_{ij} is the stress component (i,j =1, 2 or 3), u_i is the displacement and Γ is the perimeter or path of integration.

Linear cumulative damage (Bills) – failure is predicted when the cumulative damage exceeds unity [44 - 46]:

$$D = \frac{1}{(\sigma_{t0} - \sigma_{CR})^\beta} \int_0^t \frac{(\sigma_t - \sigma_{CR})}{a_T(t)} dt \geq 1 \tag{9}$$

where D = cumulative damage fraction, σ_t = applied true stress, σ_{CR} = critical true stress below which no failure occurs, σ_{t0} = true stress required to fail in time t_0 (reduced time), β = slope of time-to-failure plot and a_T is the time-temperature shift factor.

There are several advantages in using a failure criterion based on a cumulative damage model (i.e. failure depends on a critical level of stress over a finite time). The model can provide a predictive capability which takes into account the total history of loading whose effects are cumulative in time and proportional to the work done on the propellant. Also, the model can be extended to include strain-rate and temperature variations, for materials where a time/temperature equivalence assumption is valid [18].

2.3. Predictive Assessment

The two approaches used in the evaluation of structural integrity of solid rockets are (1) deterministic techniques using a number such as margin of safety to define a minimum required level and (2) probabilistic methods where structural integrity is determined by a level of reliability.

2.3.1. Deterministic Approach

Safety Factors / Margin of Safety

Margins of safety and safety factors are traditionally used in the design phase and service life assessment to account for uncertainties in the material properties characterization and structural analysis. There are various definitions used in the solid rocket community in different countries to describe safety margin, safety ratio, safety factor, and so on. These values are all measures of the excess capability of the propellant grain or bondline material over the design requirement. A value is calculated at each critical location in the motor for various loading conditions, using appropriate failure criteria.

One definition of the grain safety margin is:

$$MS = (C - S)/S \tag{10}$$

where C is the material's measured capability and S is the stress or strain calculated from the induced loads. MS is a measure of the excess propellant stress or strain capability over the design requirement and should be greater than 1 for a rocket motor to be structurally sound.

Service Life Prediction

The deterministic approach can be readily applied to service life analysis of solid rockets. The method involves identifying the likely failure mode(s) and determining whether failure will occur under the operational and storage loading conditions seen by the rocket during its service life. Usually the 'worst case' condition is used to give a conservative prediction. The change in response and failure properties of the propellant caused by ageing should also be considered in the structural and failure analyses. Fig. **15** illustrates how service life could be predicted from safety margins. For example, the end of service life is predicted when the allowable stress or strain is equal to or less than the induced values.

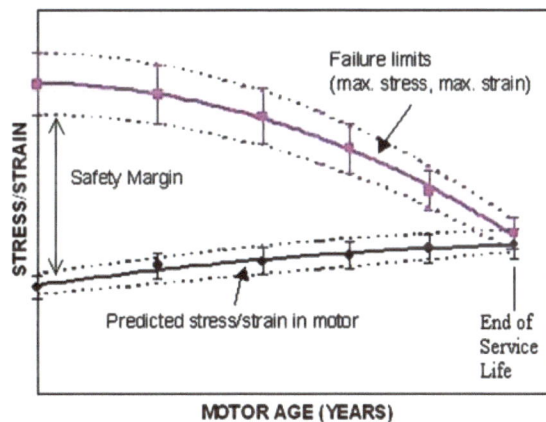

Figure 15: Service Life Projection Using Deterministic Approach.

2.3.2. Probabilistic Approach

The probabilistic methods used in the evaluation of structural integrity of solid rockets consider the statistical variability of the induced loads (e.g. environmental) and material capability. They are both treated as random variables described by probability density functions (pdf). The failure probability (P_f) is then the probability that the material strength random variable, C, is less than or equal to the loading variable, S, i.e., $P_f = P(C \leq S)$.

One approach for determining the failure probability is using the stress-strength interference method. The steps involved in the probabilistic modeling are very briefly described below (mathematical derivations have been left out for the sake of simplicity).

(1) The statistical variability of the induced load is treated as a random variable that can be described by a linear combination of sinusoidal equations [25]:

$$\sigma(t) = \sigma_m + \sigma_y \cos(\omega_y t + \phi_y) + \sigma_d \cos(\omega_d t + \phi_d) \tag{11}$$

where $\sigma_m, \sigma_y, \sigma_d$, are the mean, yearly and daily stress amplitudes; $\sigma(t)$ is the stress at time t;

ω_y, ω_d, are the seasonal/yearly and daily frequency;

$\phi_y, \phi_d,$ are the yearly and daily phases.

The stress/temperature data may be obtained from rocket motors instrumented with miniature stress sensors and thermocouples [26] or by finite element analysis. More details on rocket motor instrumentation with state-of-the-art miniature sensors are given in Chapter 5 of this book.

$$E(S) = \int_{-\infty}^{\infty} S(T) f_T(T) dT \qquad (12)$$

(2) The most appropriate failure criterion is chosen to describe the material capability. To characterise the mean and variance of the experimental data for material strength at various temperatures as a function of time, the statistical nature of the material strength is calculated by integrating the following equation:

where $E(S)$ is the expected strength, $S(T)$ is the measured strength as a function of temperature and $f_T(T)$ is the pdf for temperature. The mean and variance estimates of the strength as a function of time are then calculated.

(3) If the material capability is assumed to be Weibull distributed, the probability density function (pdf) of the strength is given by:

$$\mu_S = \beta\Gamma(1+\frac{1}{\alpha}), \quad v_S^2 = \beta^2\left[\Gamma\left(1+\frac{2}{\alpha}\right) - \Gamma^2\left(1+\frac{1}{\alpha}\right)\right] \quad f_S(S) = \frac{\alpha}{\beta}(\frac{S}{\beta})^{\alpha-1}\exp[-(\frac{S}{\beta})^{\alpha}], \quad S \geq 0 \qquad (13)$$

where the parameters α and β are obtained by numerical solution of the equations for the strength mean, μ_s, and variance, v_S.

(4) The probability of failure at any time, $p_f(t)$, is defined by the stress-strength integral below:

$$p_f(t) = \int_0^\infty\int_0^y f(f_S(S,t)dS \, f_y(y,t)dy \qquad (14)$$

where $f_S(S,t)dS = pdf$ of Weibull distribution for strength random variable and $f_y(y,t)dy = pdf$ of Weibull distribution for induced stress random variable.

(5) The analysis can be further refined by incorporating cumulative damage and ageing models.

The application of probabilistic methodologies to evaluate the structural reliability of solid rockets requires a large database to describe the statistical models adequately. It is not yet a widely used method for determining the service life of solid rockets. The accuracy of this method depends on the accuracy of the stress / strain and temperature data collected from instrumented rocket motors and assume that the statistical distributions (e.g. Weibull) used are valid.

3. VULNERABILITY ASSESSMENT

A major emphasis in solid rocket technology is to lower the vulnerability of the propellant grain to impact loading (such as fragment / bullet attack, accidental dropping) without compromising ballistic performance. With the introduction of Insensitive Munitions (IM) policies [1] in many countries (e.g. NATO nations, Australia), it is mandatory for new rocket systems to pass IM tests specified in MIL-STD-2105B before entering into service. These include bullet impact and fragment impact tests [29]. Both these tests and small-scale projectile impact tests show that the high strain-rate mechanical properties of the propellant are important to the ease of ignition and violence of the deflagration reaction [1, 31-34]. Hence, a fundamental understanding and characterisation of the high strain-rate mechanical behaviour of propellants is important in order to (1) reduce the vulnerability of rocket systems to impact loading and (2) determine the structural integrity of the rocket, to meet the design and IM requirements.

High-Strain Rate Impact Tests

Small-scale laboratory tests for assessing the response of rocket component materials (such as the propellant) to high velocity projectile impact are often used during the early stage development of rocket systems to determine / verify the sensitivity and mechanical behaviour of the component materials, because they are relatively simple and inexpensive compared to full-scale testing. One such small-scale testing method that offers the promise of being adopted as a standard test to study the impact ignition phenomena [1] is the modified Hopkinson Bar [8, 18, 31-33]. It has been used for (1) characterizing the high strain-rate mechanical properties required for input into structural integrity analysis of the rocket under dynamic (impact) loading conditions, (2) parametric studies during the early stage design of energetic materials and (3) fundamental studies to understand the deformation, fracture and ignition events during projectile impact.

The split Hopkinson Bar has been widely used for dynamic testing of solid materials [35-38]. Variations of the split Hopkinson Bar described by Kolsky [35] and Hauser [36] are the most commonly used methods. It consists of two long elastic bars, between which the specimen is inserted. The stress-strain history of the specimen can be indirectly determined by measuring the axial strain history in both the input and output bars and applying some simplifying assumptions. For example, in the design of the experiment, the length-to-diameter ratio of the elastic bar must be large to achieve a reasonable approximation to uniaxial stress wave propagation. Since one dimensional elastic wave propagation theory is used in analyzing the data, the split Hopkinson Bar technique is limited by the elastic limit of the input bar, which restricts both the amount of strain and strain-rate that can be obtained.

The modified Hopkinson Bar [40] originally developed by Wulf and Richardson at the Defence Standards Laboratories in NSW, Australia (now DSTO) is a method which is capable of measuring the stress-strain relationship of materials in compression over a much wider range of strain and strain-rates. In this method, the normal input bar was removed and a hardened non-deforming projectile allowed to impact the specimen directly. The moving input bar (projectile) was fired at a specimen fixed to the stationary output bar using a ramset gun. Specimen strain was measured independently from the position of the back end of the projectile using a coaxial capacitor. Depending on the projectile velocity and specimen size, the testing range can be extended to true strains of up to 2.0 and strain-rates from 'quasi-static' to 10^5 s^{-1}.

Fig. **16** shows a schematic diagram of the modified Hopkinson Bar used for assessing the response of rocket propellants to high velocity projectile impact. Instrumentation such as fast response photodetectors, blast pressure gauges, and high speed photography [18, 31-33] were used in conjunction with the modified Hopkinson Bar apparatus to study the deformation, fracture and ignition processes during the projectile impact event.

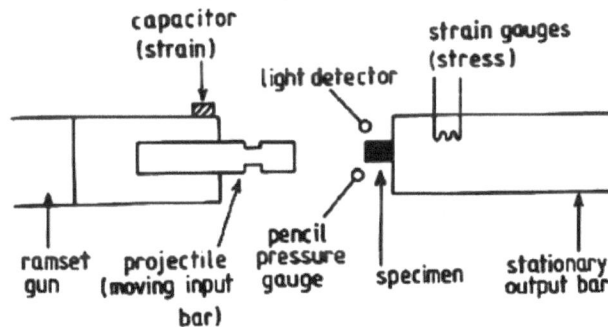

Figure 16: Schematic diagram of modified Hopkinson Bar test, as taken from reference [32].

Deformation and Fracture Processes

Studies on high velocity projectile impact of solid propellants and sub-scale rocket motors showed that the impact sensitivity and mechanism are governed by the high strain-rate mechanical properties of the propellant [31-34], which influence the failure mode (e.g. ductile and /or shear failure, brittle fracture, elastic deformation). Typical stress-strain curves from the modified Hopkinson Bar test (strain-rate ca. 10^3s^{-1}) for three solid propellants exhibiting different high strain-rate mechanical properties are illustrated in Fig. **17**.

The stress-strain data show that the rubbery composite propellant B undergo extensive plastic deformation before ductile and shear failure (evident from visual examination of the impacted specimens after testing). On the other hand, the stress-strain curve of propellant C is characteristic of elastomeric polymers, undergoing quasi-rubber like deformation. This propellant is mechanically weak and exhibits the low load capability and large strain viscoelastic behaviour characteristic of elastomers. Examination of the specimen after testing showed that localized heating had occurred, as a result of the high temperature produced in the regions of maximum shear stress during compression. Propellant A undergoes brittle failure and shows comparatively less plastic deformation, and has the highest maximum stress, modulus and fracture toughness but a low strain to failure.

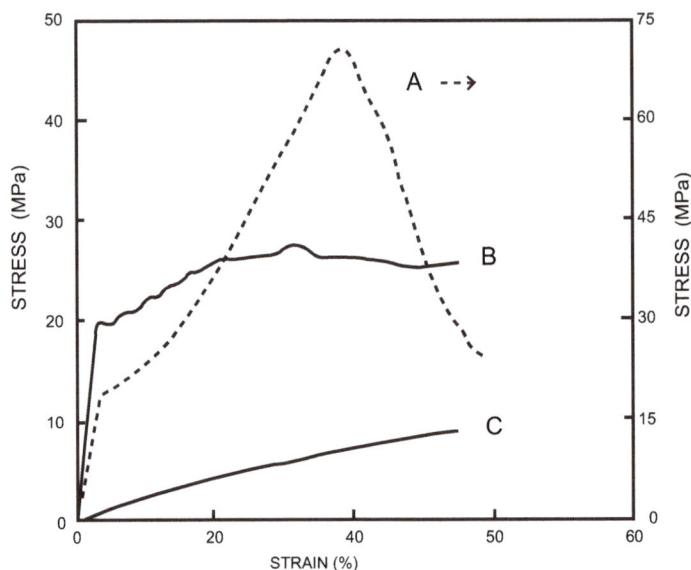

Figure 17: Stress-strain curves, at ambient temperature, of various propellants with specimen diameter-to-length ratio of 10:8, showing: A - brittle, B- ductile and C – elastomeric mechanical behaviour, as taken from reference [33].

Photographs taken at various times (at intervals of 25-50µs) during the impact event [33] show the deformation, fracture and ignition processes during projectile impact and provide clear evidence of the role of plastic deformation and adiabatic shear in the ignition process of a typical composite propellant (see Fig. **18**).

The sequence of events, photographed during projectile impact, leading to fracture and ignition of the ductile propellant B are shown in Figs. 18a-i. Compression of the propellant on impact results in viscoelastic / plastic deformation and flow, and crack initiation and propagation (Figs. 18a-e). The first signs of ignition are observed at the specimen-projectile interface (region where the shear stress is maximum) and around cracks (Fig. **18**f). Crushing of the remaining propellant to a thin layer, fragmentation, and further ignition and initial burning of the propellant occur in Fig. **18**g. For this propellant, the photographs provide clear evidence for the role of plastic deformation and adiabatic shear in the impact ignition process.

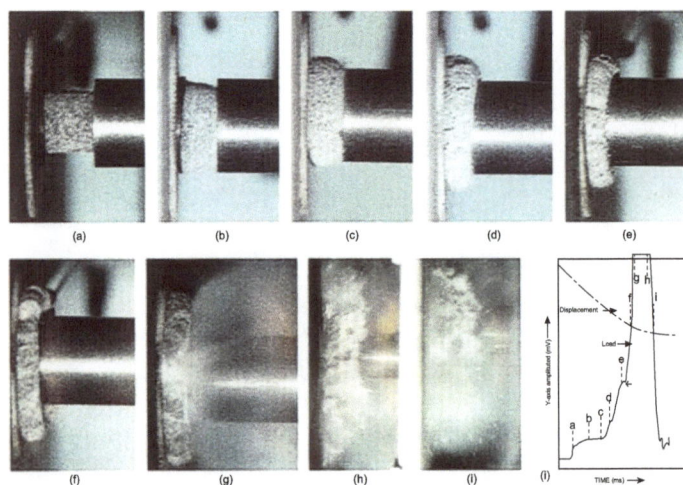

Figure 18a-i: Photographs show the sequence of events leading to fracture and ignition of propellant B, as taken from reference [33]. Vertical lines on corresponding load-time curve indicate when photographs were taken.

The deformation and ignition mechanisms of propellant C are illustrated in Fig. **19**a-i. The photographs (Figs. **19**a-d) show the propellant undergoing large-strain viscoelastic / plastic deformation with little or no cracking, typical of the elastomeric deformation charactersitics suggested by the stress-strain curve. There is some evidence of ignition starting at the specimen-projectile interface (Fig. **19**e). In Fig. **19**f, two very distinct phases, a solid layer and a white translucent rubbery-viscous ('melt') layer, and jetting are evident. Ignition is observed in the rubbery viscous layer near the junction of the solid and molten layers, where the shear stress is expected to be a maximum, suggesting that ignition is by deformation work of the viscous layer. Ignition is confined in the molten layer, providing clear evidence that plastic deformation in the solid phase does not generate enough heat to ignite this propellant.

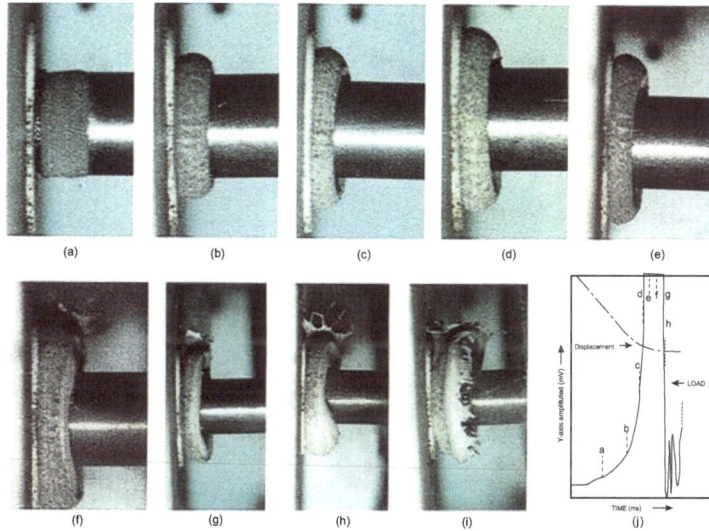

Figure 19: Photographs show the sequence of events leading to fracture and ignition of propellant C, as taken from reference [33]. Vertical lines on corresponding load-time curve indicate when photographs were taken.

The fracture and ignition processes for propellant A are shown in Figs. 20a-g. This propellant exhibits brittle fracture behaviour under impact loading at high strain-rate. The stress-strain curves and photographs show that this propellant undergoes relatively less plastic deformation and flow before fracture (Figs. 20a-d). Ignition occurs after brittle fracture and is first visible around the grains of the propellant where cracking had occurred (figs. 20e-f). These observations suggest that ignition by brittle fracture, as a result of the heat released during fast crack propagation, is a possible heating mechanism.

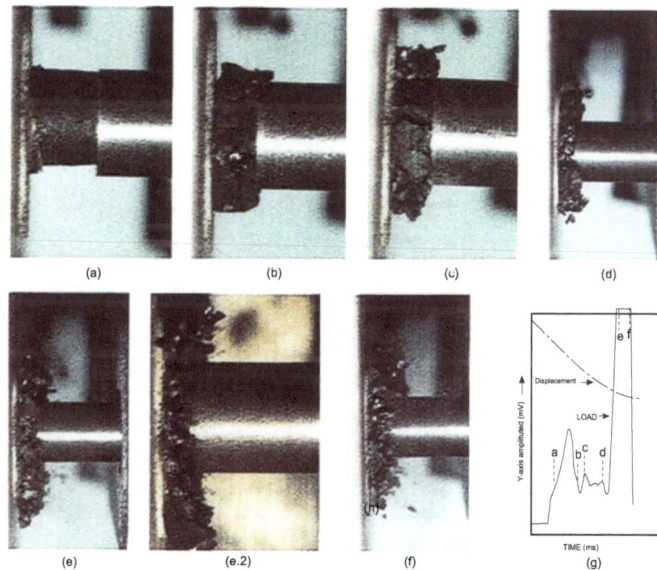

Figure 20: Photographs show the sequence of events leading to fracture and ignition of propellant A, as taken from reference [33]. Vertical lines on corresponding load-time curve indicate when photographs were taken.

Temperature Dependence of High Strain-Rate Mechanical Properties

In service, solid rockets are often required to operate over a wide range of temperature which can cause significant changes in the physical and mechanical properties of the propellant. It is well known that for some solid propellants, the most important change in impact strength occurs over a relatively narrow temperature range where the type of fracture changes between brittle and ductile [31]. Hence, it is important to determine the temperature dependence of the mechanical properties over a wide range of strain rates.

The load-displacement plots at various temperatures, measured in the strain-rate range 10^3 to 10^4 s^{-1} are illustrated in Figs. 21 and 22 for propellants A and B respectively. They show that for propellant A, brittle fracture occurred at around -40°C, but propellant B remained ductile even at -60°C. The brittle-ductile transition temperature seen in the stress-strain data of propellant A is related to the glass transition temperature determined from dynamic mechanical thermal analysis [31].

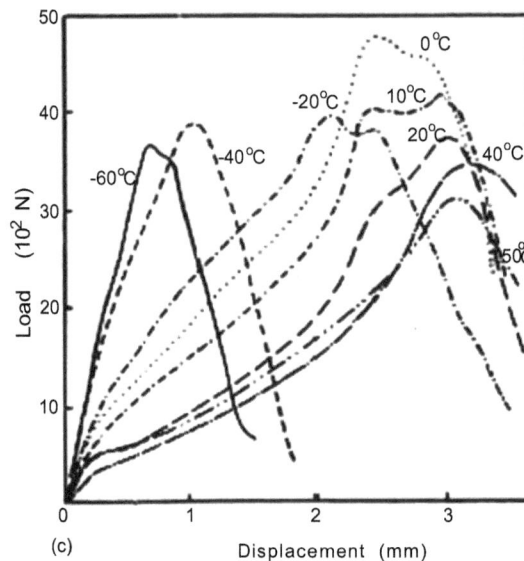

Figure 21: Load-displacement at various temperatures for propellant A from modified Hopkinson Bar test, as taken from reference [31].

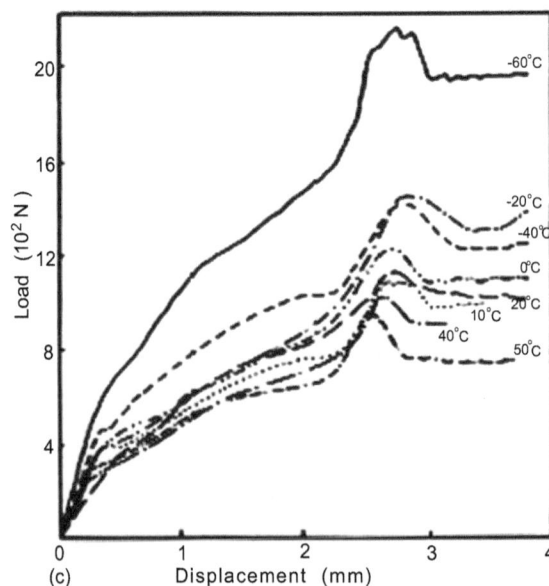

Figure 22: Load-displacement at various temperatures for propellant B from modified Hopkinson Bar test, as taken from reference [31].

Our studies [41-42] showed that there is a direct correlation between impact ignition sensitivity and fracture toughness of the propellant. Fig. **23** shows a plot of fracture toughness vs. temperature for propellants A and B. Below the brittle-ductile transition temperature of propellant A, the total fracture toughness (elastic + post yield ductile regions of stress-strain curve) increases with increasing temperature as the compliance of the material increases. Propellant B is ductile and has a higher fracture toughness. Above the transition temperature (> -40°C) however, propellant A has a higher fracture toughness than propellant B and is much harder to ignite.

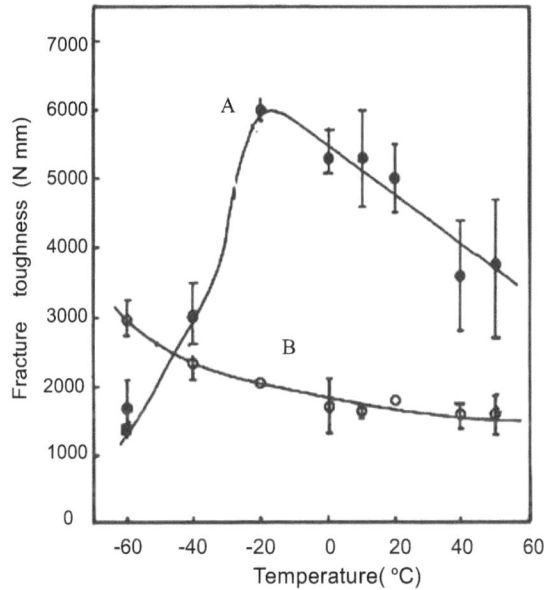

Figure 23: Total fracture toughness (elastic + post yield ductile region of stress-strain curve) vs. temperature for propellants A and B, at strain-rate $10^3 - 10^4$ s^{-1}.

Constitutive models can be developed from the stress-strain data at various temperatures using the procedures described earlier in Section 2.1.2.1. In reference [8], the stress-strain data were fitted to multivariate functions using a non-linear least-squares fitting technique. At a given temperature, the viscoelastic constitutive equation has the form:

$$\sigma = E\varepsilon^m + \eta \left| \frac{d\varepsilon}{dt} \right|^n \qquad (15)$$

where σ is in MPa, ε is in percent, and $\left| d\varepsilon/dt \right|$ is in s^{-1} (calculated from the true strain and time t). The coefficients m and n in the fitted function can be interpreted as the exponent for strain softening / hardening and the exponent for strain-rate dependence respectively. The first term in this equation models the elastic / plastic deformation. The second term is the viscoelastic or strain-rate effect, which raises the stress level for a given strain by a factor determined by the strain-rate sensitivity of the material.

For propellant B, plots of log E vs. strain-rate at different temperatures (in the range -20 to 60°C) gave straight lines that are approximately parallel, indicating a composite or master curve at a reference temperature can be used to represent the data at the other temperatures. A master curve of log E vs. log($\dot{\varepsilon} a_T$), obtained by horizontally superposing the modulus data onto a composite curve using 25°C as the reference temperature is illustrated in Fig. **7**. Here $\dot{\varepsilon} a_T$ is the reduced strain-rate and a_T is the shift factor. Constitutive models with temperature and strain-rate dependence can then be developed as described in Section 2.1.2.1.

4. SUMMARY

The structural integrity of the propellant grain and the bond systems is critical to the successful operational performance and safety requirements of the solid rocket. Accurate assessment of grain structural integrity remains a challenging area because of the uncertainties in the definition and modeling of environmental loads and the requirement for constitutive equations that incorporate damage as the source of material non-linearity. In particular, when SIA is used to predict the service life of solid rockets, it is essential that realistic specification of storage and operational environments are available and the constitutive laws are capable of correctly modeling non-linear viscoelastic behaviour, combined loads and multi-axiality.

In the future, the probabilistic approach is likely to be the preferred method for assessing structural integrity and service life prediction, as it takes into account the statistical variability of the physical, mechanical and chemical properties of the propellant and deployment temperature of the rocket system. This method, however, requires real-time data of the in-service loads (e.g. temperature, vibration), realistic ageing and cumulative damage models, and so on, to give meaningful and reliable predictions.

Structural integrity assessment of solid rockets under impact loading conditions is likely to be more important in the future as a consequence of insensitive munitions (IM) and more stringent safety requirements. The impact sensitivity of the rocket system is related to the high strain-rate mechanical behaviour of the solid propellant. To accurately predict the vulnerability of the rocket system to projectile impact, it is necessary to develop high strain-rate constitutive models that can take into account temperature and strain-rate dependence and incorporates damage.

5. REFERENCES

[1]   Victor AC. Insensitive Munitions Technology, Chapter 9 in "Tactical Missile Propulsion", Edited by Jensen GE, Netzer DW, Progress in Astronautics and Aeronautics 1996; 170: 273-361.

[2]   Francis, EC, Thompson RE. Non-linear Structural Modeling of Solid Propellants. Proceedings of the AIAA/SAE/ASME Joint Propulsion Conference (Cincinnati, OH), AIAA, New York 1984; 1-5.

[3]   Francis EC. Propellant Nonlinear Constitutive Theory Extension. United Technologies, Chemical Systems Division, Report No. AFRPL-TR-83-071 1984.

[4]   Schapery RA. A Micromechanical Model for Nonlinear Viscoelastic Behaviour of Particle-Reinforced Rubber with Distributed Damage. Engineering Fracture Mechanics 1986; 25: 845-867.

[5]   Schapery RA. A Theory of Mechanical Behaviour of Viscoelastic Composites with Growing Damage and Other Changes in Structure.J Mech Physics Solids 1990; 38: 215-253.

[6]   Lee YT. Coupled Thermomechanical Effects in High Solids Propellant. AIAA J 1979; 17: 1015-1017.

[7]   Ho SY, Carè G. Modified Fracture Mechanics Approach in Structural Analysis of Solid Rocket Motors. Journal of Propulsion and Power 1998; 14(4):409-415.

[8]   Ho SY. High Strain-Rate Constitutive Models for Solid Rocket Propellants. Journal of Propulsion and Power 2002; 18(5): 1106-1111.

[9]   Wong FC. On the Prediction of Mechanical Behaviour of Particulate Composites Using an Improved Mori-Tanaka Method. Defence Research Establishment Valcartier, Report No. DREV-R-9514 1997.

[10]   Wong FC. Analysis of Particulate Composite Behaviour Based on Nonlinear Elasticity and an Improved Mori-Tanaka Theory. Defence Research Establishment Valcartier, Report No. DREV-R-9815 1998.

[11]   Finite Element Handbook. Editor-in-Chief, Kardestuncer H and Project Editor, Norrie DH. McGraw-Hill Book Company 1987.

[12]   Zienkiewicz OC, Taylor RL. The Finite Element Method. McGraw-Hill 1989; Vol 1-2.

[13]   Cook RD, Malkus DS, Plesha ME. Concepts and Applications of Finite Element Analysis.John Wiley, 3rd Edition 1989.

[14]   Mooney M. A Theory of Large Elastic Deformation. Journal of Applied Physics 1940; 11:582-592.

[15]   Rivlin RS, Saunders DW. Large Elastic Deformations of Isotropic Materials (VII. Experinebts on the Deformation of Rubber). Phil. Trans. A. 1951; 243:251-265.

[16]   Odgen RW. Large Deformation Isotropic Elasticity – On the Correlation of Theory and Experiment for Incompressible Rubberlike Solids.Rubber Chemistry and Technology 1973; 46: 398-416.

[17]   Swanson SR, Christensen LW. A Constitutive Formulation for High Elongation Propellants. J. Spacecraft 1983; 20: 559-566.

[18]   Ho SY. High Strain-Rate Impact Studies of Pre-damaged Rocket Propellants, I. Characterisation of Damage Using a Cumulative Damage Failure Criterion. Combustion and Flame 1996; 104:524-534.

[19]   Simo JC. On a Fully Three-Dimensional Finite-Strain Viscoelastic Damage Model: Formulation and Computational Aspects. Computer Methods in Applied Mechanics and Engineering 1987; 60: 153-173.

[20]   Kinloch AJ, Tod DA.New Approach to Crack Growth in Rubbery Composite Propellants. Propellants, Explosives, Pyrotechnics 1984; 9(2): 48-55.

[21] Ho SY. Viscoelastic Response of Solid Rocket Motor Components for Service Life Assessment. J. Mat. Sci. 1997; 32(19): 5155-5161.

[22] Ho SY, Tod DA. Mechanical Failure Analysis of Rubbery Composite Propellants Using a Modified Fracture Mechanics Approach. Proceedings of the 21st International Conference of ICT (Karlesruhe, Germany) 1990; Paper 30:1-14.

[23] Ho SY. Viscoelastic Response of Aged Rubbery Composite Propellants. Proceedings of the 16th Meeting of The Technical Cooperation Program, Subgroup W, Technical Panel 4, 1991.

[24] Kinloch AJ, Young RJ. Fracture Behaviour of Polymers, Elsevier Publisher, London 1983.

[25] Smith DH, Ho SY. Analysis of Measured Stress Data in a Solid Rocket Motor. ANZIAM J 2000; 42(E): C1258.

[26] Ho SY, Macdowell. New Service Life Methodologies for Solid Propellant Rocket Motors. DSTO Research Report, Aeronautical and Maritime Research Laboratory, DSTO-RR-0099, 1997; AR-010-186.

[27] Ho SY. Finite Element Vibration Analysis of the Nulka Round. Paper presented at the Joint US / Australia Nulka Decoy Technical Meeting, Australia 2001.

[28] Ho SY, Bateup ML. Life Extension of Crew Module Rocket. DSTO Research Report, Aeronautical and Maritime Research Laboratory, DSTO-RR-0183, 2000; AR-011-532.

[29] Hazard Assessment Tests for Non-Nuclear Ordnance. Military Standard, MIL-STD-2105B, 1994.

[31] Ho SY, Fong CW. Temperature dependence of High Strain-Rate Impact Fracture Behaviour in Highly Filled Polymeric Composite and Plasticized Thermoplastic Propellants. Journal of Material Science 1987; 22:3023-3031.

[32] Ho SY, Fong CW, Hamshere BL. Assessment of the Response of Rocket Propellants to High-Velocity Projectile Impact Using Small-Scale Laboratory Tests. Combustion and Flame 1989; 77:395-404.

[33] Ho SY. Impact Ignition Mechanisms of Rocket Propellants. Combustion and Flame 1992; 91:131-142.

[34] Manners DJ. Model Rocket Motor Studies for Reduced Vulnerability. Proceedings of the 19th International Annual Conference of ICT, Karlesruhe 1988; paper 29:1-14.

[35] Kolsky H. An Investigation of the Mechanical Properties of Materials at Very High Rates of Loading. Proceedings of the Physical Society 1949; 62:676-700.

[36] Hauser FE. Techniques For Measuring Stress-Strain Relations at High Strain-Rates. Experimental Mechanics 1966; 6(8)395-402.

[37] Davies E.D.H. Hunter S.C. The Dynamic Compression Testing of Solids By The Method of the Split Hopkinson Pressure Bar. J. Mech. Phys. Solids 1963; 11: 155-179.

[38] Bertholf LD. Feasibility of Two-Dimensional Numerical Analysis of the Split-Hopkinson Pressure Bar System. Journal of Applied Mechanics 1974; March Issue: 137-144.

[39]Follansbee PS, Frantz C. Wave Propagation in the Split Hopkinson Pressure Bar. Journal of Engineering Materials and Technology 1983; 105:61-66.

[40] Wulf GL, Richardson GT. The Measurement of Dynamic Stress-Strain Relationships at Very High Strains. Journal of Physics E: Scientific Instruments 1974; 7:167-169.

[41] Ho SY. The Mechanism of Impact Ignition of Energetic Materials. Proceedings of the Ninth Symposium (International) on Detonation (Portlant, OR) 1989.

[42] Ho SY, Fong CW. Relationship Between Impact Ignition Sensitivity and Kinetics of the Thermal Decomposition of Solid Propellants. Combustion and Flame 1989; 75:139-151.

[43] Ho SY. Modelling Cook-off Reaction Violence of Confined Energetic Materials. Chapter 2.15 in Solid Propellant Combustion and Motor Interior Ballistics. Progress in Astronautics and Aeronautics, AIAA Series 185, 2000.

[44] Miner MA. Cumulative Damage in Fatigue. Journal of Applied Mechanics 1945; 12:A159-A164.

[45] Duerr TH, Marsh BP. Solid Propellant Grain Structural Behaviour, Chapter 5 in "Tactical Missile Propulsion", Edited by Jensen GE, Netzer DW, Progress in Astronautics and Aeronautics 1996; 170: 273-361.

[46] CPIA Publication 21. ICRPG Solid Propellant Mechanical Behaviour Manual, 1965; Section 4.3.3.

INDEX